Patterns
in the Sky

AN INTRODUCTION TO STARGAZING

For my father, and in memory of my late mother;
without their encouragement
my celestial journey could not have begun.

And for Lynda, who has explored
all the constellations with me.

OTHER BOOKS IN THIS SERIES

Secrets of Stargazing by Becky Ramatowski
Exploring the Moon by Gary Seronik

© 2006 New Track Media LLC
Sky Publishing
90 Sherman Street
Cambridge, MA 02140-3264, USA
SkyTonight.com

Library of Congress Cataloging-in-Publication Data

Hewitt-White, Ken, 1951-
  Patterns in the sky : an introduction to stargazing / Ken Hewitt-White.
    p. cm. -- (NightSky astronomy for everyone series)
  Includes index.
  ISBN 1-931559-39-2 (pbk.)
  1. Stars--Amateurs' manuals 2. Constellations--Amateurs' manuals 3. Stars--Observer's manu-
als. 4. Constellations--Observer's manuals. I. Title.
QB63.H49 2006
523.8--dc22
                          2006032883   Printed in Canada

Patterns
in the Sky

AN INTRODUCTION TO STARGAZING

Ken Hewitt-White

SKY PUBLISHING
A New Track Media Company
Cambridge, Massachusetts

Table *of* Contents

►Introduction

Every clear evening at nightfall, a silent picture show materializes overhead. It was dreamt up thousands of years ago by Mediterranean and Middle Eastern peoples who projected their deep-rooted myths and legends onto a starry stage. Their epic stories featured a mind-boggling variety of superhuman heroes, exotic wild beasts, and horrific monsters. We know them as the constellations.

The main sky patterns have been passed down to us from a catalog of 48 constellations compiled by the Greek astronomer Ptolemy in AD 150. Ptolemy's list slowly expanded over the centuries to the 88 constellations we recognize today. Some of the additions fill gaps between the Greek figures, while others employ "new" stars that were sighted by seagoing navigators during their explorations of the Southern Hemisphere.

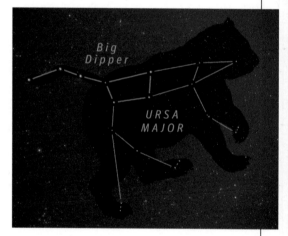

The ancient sky figures — Orion, Leo, Andromeda, Cygnus, and many others — are still visible (at least in part) from your own backyard. This book presents 45 of them grouped according to the seasons in which they appear. With each pattern you'll find a basic star chart, a non-technical description of interesting stars or other objects (all plotted on the chart), and one or two short legends — mostly from classical mythology, but a few from other sources as well.

The emphasis in the coming pages is on stargazing without telescopes. You won't need any special gear except, perhaps, binoculars if you have them, plus a dim, red-light flashlight to preserve your night vision while reading the charts. (If you don't have an observer's light, wrap red cellophane around the front end of a flashlight.) Getting "dark adapted" is an essential part of any night sky observing session. You'll see fainter stars if the pupils of your eyes are fully open. This means avoid staring at computer and tele-

STARS UNDER SIEGE: In the battle between streetlight and starlight, the streetlights usually win — but not always, as these exposures of the Milky Way from a location looking south over Toronto, Canada, attest. The image on the left was shot during a power failure on August 14, 2003. The picture on the right was taken once power was restored one night later.

vision screens for at least 10 minutes before heading outside. Once outdoors, stay away from porch lights and street lamps as much as possible. With stargazing, darkness is everything.

It is tragic, then, that the night sky is not as dark as it used to be. The relentless spread of urban lighting has obliterated the glorious, star-filled nights of yesteryear. Those of us who live in big cities can't pick out the dimmer stars or trace the subtle glow of the Milky Way. Despite this limitation, the steady sales of stargazing books and magazines (not to mention telescopes) confirm that people's interest in the night sky is stronger than ever. Although my descriptions of the constellations emphasize what can be observed from a suburban environment, I encourage you to seek out darker observing sites whenever you can.

A crystal-clear night sky resplendent with stars seems almost magical. This is appropriate because the constellations themselves are a grand illusion. Most of the stars in a constellation have no physical connection with each another. The stick-figure outlines that strike our imagination result mainly from chance alignments. The modern-day constellation boundaries have been set by international agreement; they exist solely to aid the locating and naming of sky objects. In an era of astrophysics and giant telescopes, the constellations have become an astronomical anachronism.

But that doesn't lessen their appeal. The constellations are a link across the barriers of time, a beautiful gift from our half-forgotten ancestors. Once you become familiar with the constellations, they'll be your companions for life. Let this book be your guide.

Patterns in the Sky
2

▶Capsule Cosmos

The night is crisp and clear, and you're out for a stroll. Overhead, the darkness is pierced by hundreds of twinkling jewels. All the stars you can see belong to just a portion of our Milky Way Galaxy. How big is the Milky Way? What lies beyond? Join me on a quick tour of the universe-at-large to see where the Earth fits in the cosmic scheme of things.

Begin by thinking big. The universe is an immense place full of huge, physical structures called *galaxies*. Galaxies can be categorized into several main types, the most photogenic of which are the *spiral galaxies*. A typical spiral contains lots of stars, gas, and dust all bound together by gravity in a pinwheel-shaped disk encircling a central hub or bulge. Galaxies of every type congregate in *galaxy clusters*, each one containing thousands of members. The galaxy clusters, in turn, tend to form *superclusters*. Astronomers have detected galaxies out to a distance in excess of 10 billion light-years.

The nearest galaxy cluster to us — a mere 60 million light-years away — is the Virgo Galaxy Cluster, which is part of a vast archipel-

FAINT FUZZIES: The fuzzy-looking objects in this photograph are galaxies near the center of the Virgo Galaxy Cluster. Each cotton ball of light represents the combined energy output of hundreds of billions of stars.

ago of galaxies called the Virgo Supercluster. Along the outskirts of the Virgo Supercluster is a small clump of galaxies known as the Local Group. Two massive spiral galaxies dominate the Local Group. One is the famous Andromeda Galaxy (page 79), which at a distance of 2.5 million light-years, is the nearest major galaxy to Earth. The other big player is the Milky Way itself — *our* galaxy.

INSIDE THE MILKY WAY

The Milky Way stretches some 100,000 light-years from tip to tip and contains upwards of 400 billion stars. Its central bulge is packed with ancient suns more than 10 billion years old while its spiral disk is dominated by mostly younger stars. Young stars tend to form in gravitationally bound families called *open clusters*. A spectacular example is the Pleiades Cluster (page 27), about 400 light-years away in the constellation Taurus. Much denser *globular clusters* congregate in a halo around the galaxy's central bulge. Globular clusters are the senior citizens of the Milky Way, for they are as old as the galaxy itself. A superb example is the Great Sagittarius Cluster (page 65), a swarm of aging suns 10,000 light-years from Earth.

STELLAR NURSERY: A crimson-colored cloud of hydrogen gas, the Orion Nebula cradles newborn suns.

Amid all the gas and dust that permeates the spiral arms of the Milky Way are gigantic, luminous clouds called *emission nebulas*. Some of these nebulas are stellar birthplaces. The Great Nebula in Orion (page 25), a brilliant cloud of hydrogen 1,500 light-years

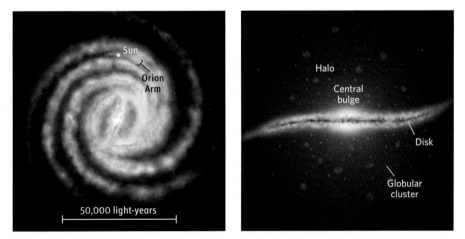

MILKY WAY METROPOLIS: Our Milky Way Galaxy is a disk-shaped city of stars. The Sun resides in the disk, about three-fifths of the way from the center to the edge, in a region called the Orion Arm.

Patterns in the Sky

from Earth, is illuminated by the compact cluster of young stars that formed inside it. Astronomers think that our Sun and its family of planets formed in a gaseous cocoon much like the Orion Nebula 4.6 billion years ago.

PLANETS

Our solar system consists of eight "classical" planets, several "dwarf" planets (including Pluto, a moon-sized world of ice and rock roughly six billion kilometers away), more than 100 moons, thousands of rocky asteroids, and thousands more icy comets. Dominating the solar system are four huge atmospheric planets — Jupiter, Saturn, Uranus, and Neptune. Plying huge orbits hundreds of millions of kilometers from the Sun, these gas giants are hot on the inside but frosty on the outside. Their many moons have surfaces of ice and rock, but the gas giants have no solid surfaces at all.

Four small, rocky worlds orbit much nearer the Sun. The surface environments of these *terrestrial planets* are radically different from each other. The Sun's nearest planet, Mercury, is an airless world scarred with craters. The second planet, Venus, is a desiccated, volcanic broiler shrouded in a thick, poisonous atmosphere. By contrast, the fourth planet, Mars, hints at a watery past but retains only a freeze-dried landscape surmounted by an exceedingly thin atmosphere. The planet between Venus and Mars — *our* planet — is much more inviting. The Earth's nitrogen-oxygen atmosphere protects a warm, wet surface teeming with life. Most remarkable of all, the Earth's biosphere has produced a singular lifeform that has the remarkable, wondrous ability to observe the universe around it.

LOOKING UP

Galaxies, stars, planets . . . the basic layout of the universe seems fairly straightforward. But when you gaze upward on a clear night, you don't sense much cosmic structure. All you see is a maze of pinpoints scattered across the black dome of the sky. Are any of them planets? Which one is the North Star? Where do I look to see Orion? When is the best time to admire the Milky Way? You'll soon be able to answer those questions — and others you haven't even thought of yet — with the aid of this book. However, to get the most out of the sky show, you should have a basic understanding of how the sky itself works. The next chapter takes us "backstage" to learn the essential structure of that black dome above us.

► Never Fear the Celestial Sphere

We all know that the Earth rotates on an axis and that it revolves around the Sun. We also know that the stars lie at vastly differing distances and have no physical connection with Earth. Yet when we go stargazing, our senses suggest something different. The stars appear as though they're attached to the inside of a huge, invisible sphere that turns around a motionless Earth. This imaginary structure is called the *celestial sphere* and it can help you learn the night sky.

A SPHERE OF STARS: The Earth rotates, making the celestial sphere seem to turn in the opposite direction. No wonder ancient peoples thought that the Earth was the center of the universe.

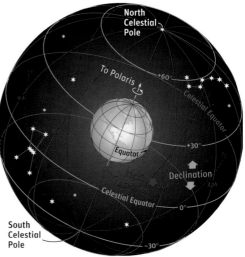

THE SKY GRID: The celestial sphere has an equator plus north and south poles, just like the Earth.

CELESTIAL SPHERE 101

The celestial sphere is marked by a coordinate system similar to that found on Earth maps. Just as geographers use latitude and longitude to identify places on a terrestrial globe, astronomers use *declination* (latitude) and *right ascension* (longitude) to locate sky objects on the celestial sphere. The Earth's equator (latitude zero) projected onto the celestial sphere becomes the *celestial equator* (declination zero). The celestial equator divides the starry sphere into northern and southern halves. An imaginary line extended along the Earth's axis of rotation through the North Pole (latitude 90° north) points to the *north celestial pole* at 90° north declination. Extending the line in the opposite direction, past the Earth's South Pole, points to the *south celestial pole* at 90° south declination.

We can't see all of the celestial sphere at any one time because the Earth itself is in the way. We perceive only the half of the sphere that's above the horizon. The half-sphere is essentially an upside-down bowl of stars suspended over a flat plane. The *horizon* circumscribing that plane is the rim of the bowl. The *zenith* is at the top of the bowl, 90° above the horizon. An imaginary line called the *meridian* passes from the northern horizon through the zenith to the southern horizon, dividing the bowl into eastern and western halves. Any sky object straddling the meridian has reached its highest point, or *culmination*. Later, we'll see that some sky objects culminate much higher than others.

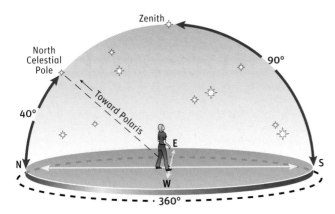

A BOWL OF STARS:
The stars appear fixed to an upside-down bowl above your observing site. The bowl measures 360° around the horizon and 90° from the horizon to the zenith. This diagram depicts the view from 40° north latitude; the north celestial pole is 40° above the north horizon.

SKY MEASURES

Here's a "handy" way to measure distances between objects on the celestial sphere. Fully extend your arm with your fingers spread as wide as possible. From the tip of your thumb to the tip of your little finger covers about 20° of sky. Your clenched fist with your thumb stuck out spans roughly 15°, while your fist alone is 10° wide. Three fingers held up in a Boy Scout salute spans 5°. Your thumb is about half that wide and the end of your little finger is 1°. Believe it or not, 1° is twice the width of the full Moon. Keep these measures in mind when you look at the star charts on the constellation pages and see a scale bar in degrees.

PERPETUAL POLAR POINTER: The circumpolar sky as viewed from latitude 40° north. Note the Little Dipper and the star Polaris next to the north celestial pole. The "pointers" in the Big Dipper aim close to Polaris.

CIRCUMPOLAR STARS

From our vantage point in the mid-northern hemisphere, the star patterns near the north celestial pole are always visible. If you keep an eye on these *circumpolar constellations*, you'll discover that they never set below the horizon. The main groups in the circumpolar sky are Cepheus, Cassiopeia, Draco, Ursa Major, and Ursa Minor. Ursa Major contains the Big Dipper — the best-known star pattern in the heavens — while Ursa Minor includes the Little Dipper. All five groups are described elsewhere in this book.

The seven bright stars forming the handle and bowl of the Big

STAR TRAILS: This long-exposure photograph taken from Hawaii shows the stars as curving streaks called star trails. The view is northward and the shortest streak is Polaris.

Patterns in the Sky

Dipper are easy to spot. The two stars at the end of the bowl are called "the pointers" because they point toward Polaris, better known as the North Star. (Polaris marks the end of the handle of the Little Dipper.) No matter the time of night or season of the year, the pointers always aim at Polaris. With a declination of 89.3°, Polaris is very close to the north celestial pole. All of the other stars wheel around the celestial pole while Polaris itself stays put — well, almost. Because Polaris is situated nearly 1° off the pole, it turns in a tiny circle of its own. The bottom line is this: When you face the North Star you face due north. South is behind you, east is on your right, and west is on your left.

THE FORBIDDEN ZONE

Just as there is a portion of the heavens that we always see, there is an identical area of sky that we never see. A glance at the diagram on the right will show that the deep southern stars in this "forbidden zone" never rise above our horizon. This explains why people in the northern United States and Canada can't admire the Southern Cross, a star pattern located far below the celestial equator. However, the view improves the farther south you live. For example, skywatchers in the southern United States see Scorpius, a mid-southern constellation, higher in the sky than do northern observers. Travellers to Mexico and Hawaii can peer deep into the forbidden zone to identify a number of exotic sky treasures, including the Southern Cross (see page 89).

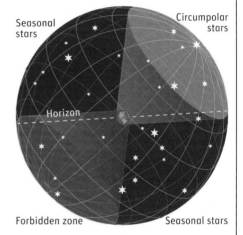

FORBIDDEN FRUIT: In this depiction of the celestial sphere for an observer at latitude 40° north, the stippled area shows the always-visible circumpolar sky. The shaded portion beneath the southern horizon indicates the identical amount of sky that is "forbidden" since it is always hidden from our view.

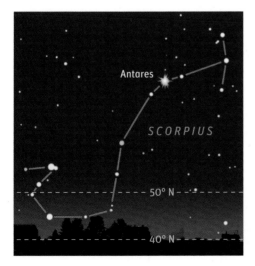

TAIL-LESS SCORPION: Scorpius is low in the southern sky on summer evenings. For observers north of about latitude 45°, a portion of the scorpion's tail is cut off by the southern horizon. Viewers farther south fare better.

Never Fear the Celestial Sphere

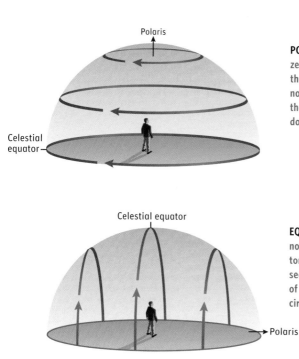

Polaris

Celestial equator

POLAR SKY: Polaris appears at the zenith and the celestial equator lies on the horizon. An observer sees all of the northern constellations, but none of the southern constellations. The stars do not rise or set.

Celestial equator

Polaris

EQUATORIAL SKY: Polaris lies on the northern horizon and the celestial equator arcs across the zenith. An observer sees half of the northern stars and half of the southern stars. There are no circumpolar (or forbidden) stars at all.

READING POLARIS

You've probably guessed by now that the amount of circumpolar sky you see depends on the latitude of your observing site. The farther north you live, the higher the north celestial pole (and Polaris) appears in your sky. In fact, the height of Polaris is an indicator of your geographic latitude. Let's check out a few locations.

If Polaris is 40° above your northern horizon, it means you live somewhere along the 40th parallel, roughly the latitude of Philadelphia and Denver. Almost half of your sky is circumpolar. If Polaris is 65° up, it means that you are at the latitude of Fairbanks, Alaska, where most of the night sky is circumpolar. If you see Polaris at the zenith, you must be at the North Pole; from latitude 90° north, *all* the stars you see are circumpolar; *none* of them rise or set.

The situation is radically different in the tropics. Viewed from Honolulu, Hawaii, Polaris is only 21° degrees above the horizon; consequently, very little of the sky is circumpolar (even the Big Dipper rises and sets). Finally, suppose you're watching the night sky from Ecuador on the Earth's equator. Polaris lies on the northern horizon and none of the stars are circumpolar. Half of the northern celestial hemisphere and half of the southern celestial hemisphere are visible at any given time. All the stars rise and set, so eventually you see every constellation in the sky.

▶ Stars for All Seasons

If the celestial sphere didn't appear to turn then learning the constellations would be easy. The stars would never change position and we'd see the same ones all the time. However, the sky sphere does turn — or, more correctly, the Earth rotates on its axis from west to east — causing the sphere to *appear* to drift from east to west, carrying the Sun, Moon, and stars with it. This motion is easily gauged at night. Pick out a bright star halfway up the southern sky at, say, 8 p.m. then observe it again an hour later. You'll notice that the star has moved somewhat to the west (that is, to your right).

If you observe the following night at 8 p.m. — exactly 24 hours later — you might expect to see the star return to the same position. In truth, the star appears slightly *west* of where it was 24 hours earlier. In fact, all the stars pass a given reference point four minutes earlier each night. A star shining just above a telephone pole at 8 p.m. on the first evening will appear over the pole at 7:56 p.m. the following night and at 7:52 p.m. the night after that. In other words, the celestial sphere completes one rotation in 23 hours, 56 minutes — not 24 hours. The reason for this is simply that the Earth is orbiting in an easterly direction around the Sun. After each 24-hour period your spot on the night side of the Earth points slightly ahead, or eastward, of where it was pointing the night before.

The four-minute difference slowly accumulates. Suppose you begin stargazing on January 1 when the winter constellations blaze most of the night. After a couple of weeks, the winter stars set almost an hour earlier than they did on January 1. By early February they're setting two hours earlier. By mid-March they're low in the west well before midnight. By the time April arrives, so have the stars of spring. After three more months have elapsed, the spring patterns have been replaced by the constellations of summer. Three months later you're enjoying the autumn groups. After a year, the Earth has completed one orbit around the Sun and the show starts all over again.

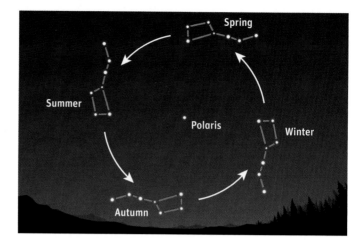

SEASONAL CLOCK:
The Big Dipper assumes different positions at nightfall depending on the time of year.

Associating certain constellations with certain seasons is somewhat arbitrary. Orion is considered a winter constellation but you can watch it rise in mid-summer — provided you don't mind staying up until dawn. Likewise, Leo is a spring pattern but you can watch it rise after midnight in mid-autumn. We divide the celestial sphere into seasons so we can concentrate on a given set of constellations during convenient evening hours. That's why the descriptions in this book are divided into four seasonal sections.

The exception is in the northern portion of the sky. There the circumpolar constellations are visible every clear night of the year — they merely change positions as they wheel in unbroken circles around the north celestial pole. For example, on winter evenings the Big Dipper stands on its handle in the northeast. In spring the Dipper appears upside down high overhead. In summer, it descends bowl-first in the northwest. During autumn, it lies near the northern horizon.

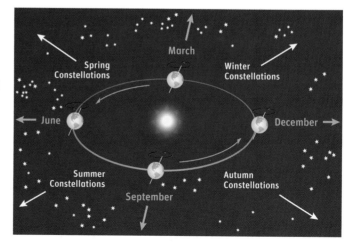

STAR SEASONS: As the Earth orbits the Sun, we look into space in different directions and therefore see different constellations. For example, on a summer evening Scorpius is visible. Six months later, that pattern is hidden from view because it's in the daytime sky.

Patterns in the Sky

USING THE SEASONAL STAR MAPS

Four circular all-sky star maps appear on foldout pages inside the front and back covers of this book. Each big chart is a representation of the evening sky in the middle of winter, spring, summer, or autumn. Using them to identify the constellations is a three-step process. Start by finding the chart for the current season. Then check the dates and times printed near the top of the page. Finally, take the chart (and a red-light flashlight) out under the night sky within an hour or so of these times.

Outside, you need to know which direction you're facing. (If you're unsure, note where the Sun sets — that's west.) Hold the chart in front of you and turn it so the label along the curved edge matching the direction you're facing is right-side up. The curved edge represents the horizon, and the stars above it on the chart now match the stars in front of you. The farther up from the chart's edge they appear, the higher up they'll be in the sky. As an example, try the winter chart inside the front cover. Hold it so that its south horizon, labelled "facing south," is right-side up. About halfway from the southern horizon to the zenith (near the center of the chart) is an hourglass-shaped pattern with three closely spaced dots slanted across its middle. This is Orion. If you go outside at the right time, face south, and look halfway up the sky, Orion will be there. Keep in mind that the stars are farther apart in the sky than they appear on the chart.

The charts are drawn for an observer near the center of North America (40° north latitude, 90° west longitude) but can be used throughout the continent and around the world at this latitude. If you're south of 40°, stars in the southern part of the sky appear higher than the chart shows (see page 9) while stars in the north are lower. If you're north of 40°, the reverse is true. Don't worry about time zones; the hours are for your *local* time.

▶ GREEK LETTERS ON STAR MAPS

On both the seasonal all-sky maps and the small star charts found throughout this book, many of the stars in each constellation are identified with Greek letters. A constellation's most brilliant star is usually called Alpha (α), the first letter in the Greek alphabet; the second brightest is Beta (β), and so on. Numerous bright stars have Arabic names that have remained in common usage. For example, while the brightest star in Leo is Alpha (α) Leonis, it is better known as Regulus.

α	Alpha	η	Eta	ν	Nu	τ	Tau
β	Beta	θ	Theta	ξ	Xi	υ	Upsilon
γ	Gamma	ι	Iota	o	Omicron	φ	Phi
δ	Delta	κ	Kappa	π	Pi	χ	Chi
ε	Epsilon	λ	Lambda	ρ	Rho	ψ	Psi
ζ	Zeta	μ	Mu	σ	Sigma	ω	Omega

Stars for All Seasons

Gemini
Cancer
Leo
Aries
Taurus ECLIPTIC Pisces
CELESTIAL EQUATOR Aquarius
Sagittarius
Scorpius
Virgo
Capricornus
Libra

THE ZODIAC ZONE:
The curving band of the zodiac straddles the celestial equator. The Sun follows the ecliptic inside the zodiac throughout the seasons.

ROLLER-COASTER ZODIAC

The zodiac is an S-shaped beltway of twelve constellations that encircles the celestial sphere. This narrow, winding zone has special significance because the Sun, Moon, and planets appear inside the zodiac and nowhere else. The Sun's seasonal path on the sky, called the *ecliptic* (the curving green line on the four all-sky maps), defines the center of the zodiac. The Sun crests high in the northern constellations Taurus and Gemini each summer then descends along the ecliptic during autumn until it bottoms out in the southern constellations Scorpius and Sagittarius every winter. The Sun climbs back up the ecliptic during spring until it reaches the top of the zodiac where the cycle begins again. Because the planets orbit the Sun in nearly the same plane, they stick close to the ecliptic as they drift up and down the zodiacal roller coaster.

The planets Mercury, Venus, Mars, Jupiter, and Saturn are visible without optical aid. An extra "star" glaring in a constellation of the zodiac is bound to be one of these five. You can be certain you're staring at a planet if: (a) the suspect shines with a steady glow instead of twinkling; (b) it creeps against the backdrop of stars over a period of nights; and (c) it's relatively bright. (Venus, Jupiter, and occasionally Mars, outshine every star in the sky.) A planet in one of the dimmer zodiacal patterns — Cancer or Pisces, for example — is often the only visible evidence of those constellations in a city sky.

PLANETARY VISITATION: The planet Mars is the brightest point of light in this photo of Taurus. The Bull's most prominent star, Aldebaran (lower-right of Mars), is no match for the red planet.

Patterns in the Sky

14

THE MAGNITUDE OF THE SITUATION

If you're a stargazing "newbie," I have some advice: Take your time. Some constellations are easier to recognize than others. Look for the most prominent ones first — not necessarily the largest, but those with bright stars. The first groupings you identify might be *asterisms*, simple patterns that contain stars belonging to one or more constellations. The Big Dipper inside Ursa Major (see page 40) is the best-known asterism in the northern sky. The Summer Triangle (page 54) is a huge asterism formed by a trio of brilliant stars.

The brightness of a star is called its *magnitude*. A simple magnitude scale was devised more than 2,000 years ago by the Greek astronomer Hipparchus. He called the brightest stars magnitude 1 and the faintest stars magnitude 6. Each step downward in magnitude represents a 2.5 times drop in brightness. A 5th-magnitude star is 100 times dimmer than a 1st-magnitude star. Fifth-magnitude stars are also about 100 times more numerous than those of 1st magnitude. Only 15 stars brighter than magnitude 1.5 are visible from latitude 40° north.

The all-sky maps in this book plot stars down to about 4th magnitude — roughly the limit of visibility for an observer in a mid-size city. Some of the individual constellation charts show slightly fainter stars. Binoculars reveal stars to 8th or 9th magnitude (depending on your location), and large observatory telescopes can see to magnitude 30 or so. At the other end of the scale, it turns out that some stars are brighter than 1st magnitude, so Hipparchus's original scheme was extended past zero into negative numbers. The brightest star, Sirius, has a magnitude of –1.4. The planet Jupiter shines at magnitude –2.7 when it's closest to Earth. Venus can be as bright as magnitude –4.6.

On the next page you'll find a table of the 45 constellations and 21 brightest stars featured in this book. Bright and dim, large and small, the constellations await you.

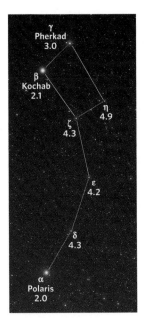

SITE TEST: Gauge the darkness of your night sky by comparing this chart of the Little Dipper with the number of stars in the Little Dipper you can actually see. Begin by locating Polaris and Kochab. Shining at 2nd magnitude, these are the Little Dipper's two brightest stars. Third-magnitude Pherkad is the third best star and should be visible in a city sky. The other four stars, glowing dimly between 4th and 5th magnitude, are tougher catches. How many Dipper dots can you spot?

▶ Meet the Cast

Here is a list of the 45 constellations featured in this book.
The page where each is described is noted after the name.

Name	Abbrev	Pronunciation	Meaning
Andromeda (79)	And	an-DROM-eh-duh	Daughter of Cassiopeia
Aquarius (86)	Aqr	a-QUAIR-ee-us	The Water Carrier
Aquila (60)	Aql	A-quill-ah	The Eagle
Aries (82)	Ari	AIR-eez	The Ram
Auriga (28)	Aur	oh-RYE-gah	The Charioteer
Boötes (46)	Boo	bo-OH-teez	The Herdsman
Cancer (43)	Cnc	KAN-ser	The Crab
Canes Venatici (44)	CVn	KAY-neez ve-NAT-ih-sigh	The Hunting Dogs
Canis Major (32)	CMa	KAY-niss MAY-jer	The Big Dog
Canis Minor (31)	CMi	KAY-niss-MY-ner	The Little Dog
Capricornus (87)	Cap	KAP-ri-KORN-us	The Sea Goat
Cassiopeia (77)	Cas	KAS-ee-oh-PEE-ah	Mother of Andromeda
Centaurus (88)	Cen	sen-TOR-us	The Centaur
Cepheus (76)	Cep	SEE-fee-us	Father of Andromeda
Cetus (85)	Cet	SEE-tus	The Sea Monster
Coma Berenices (45)	Com	KOH-mah bera-NICE-eez	Berenice's Hair
Corona Borealis (47)	CrB	kor-OH-nah bo-ree-AL-iss	The Northern Crown
Corvus (51)	Crv	CORE-vus	The Crow
Crater (51)	Crt	KRAY-ter	The Cup
Crux (89)	Cru	KRUKS	The Cross
Cygnus (58)	Cyg	SIG-nus	The Swan
Delphinus (61)	Del	del-FINE-us	The Dolphin
Draco (69)	Dra	DRAY-ko	The Dragon
Gemini (30)	Gem	GEM-in-eye	The Twins
Hercules (68)	Her	HER-cue-leez	Son of Zeus
Hydra (50)	Hya	HI-dra	The Water Snake
Leo (42)	Leo	LEE-oh	The Lion
Lepus (25)	Lep	LEE-pus	The Hare
Libra (49)	Lib	LEE-bra {LYE-bra}	The Scales
Lyra (59)	Lyr	LYE-rah	The Lyre
Ophiuchus (66)	Oph	oh-fee-YOU-kus	The Serpent Bearer
Orion (24)	Ori	oh-RYE-un	The Hunter
Pegasus (78)	Peg	PEG-uh-sus	The Winged Horse
Perseus (80)	Per	PURR-see-us	Rescuer of Andromeda
Pisces (84)	Psc	PIE-sees	The Fishes

Name	Abbrev	Pronunciation	Meaning
Piscis Austrinus (86)	PsA	PIE-sis oss-TRY-nus	The Southern Fish
Sagitta (61)	Sge	sah-JIT-ah	The Arrow
Sagittarius (64)	Sgr	saj-ih-TAIR-ee-us	The Archer
Scorpius (62)	Sco	SKOR-pee-us	The Scorpion
Serpens (67)	Ser	SER-pens	The Serpent
Taurus (26)	Tau	TOR-us	The Bull
Triangulum (83)	Tri	tri-ANG-gu-lum	The Triangle
Ursa Major (40)	UMa	UR-sah-MAY-jer	The Great Bear
Ursa Minor (41)	UMi	UR-sah-MY-ner	The Little Bear
Virgo (48)	Vir	VURR-go	The Maiden

The Bright Stars

Here are the twenty-one brightest stars described in this book. The page where each is described is noted after the name. Note that six of them — Canopus, Alpha Centauri, Hadar, Acrux, Mimosa, and Achernar — are located in the deep southern hemisphere. All are visible from the latitude of Miami, Florida.

Star	Pronunciation	Constellation	Magnitude
Sirius (32)	SEAR-ee-us	Canis Major	-1.4
Canopus (90)	can-OH-pus	Carina	-0.6
Alpha Centauri (88)	AL-fah sen-TORE-eye	Centaurus	-0.3
Arcturus (46)	ark-TOUR-us	Boötes	0.0
Vega (59)	VEE-gah	Lyra	0.0
Capella (28)	kah-PELL-ah	Auriga	0.1
Rigel (24)	RYE-jel	Orion	0.2
Procyon (31)	PRO-see-on	Canis Minor	0.4
Achernar (90)	ACHE-er-nar	Eridanus	0.5
Betelgeuse (24)	BET-el-jews	Orion	0.4 - 1.3
Hadar (88)	HAD-ar	Centaurus	0.6
Acrux (89)	A-kruks	Crux	0.7
Altair (60)	AL-tair	Aquila	0.8
Aldebaran (26)	al-DEB-uh-ran	Taurus	0.8 - 1.0
Antares (62)	an-TAIR-eez	Scorpius	0.9 - 1.2
Spica (48)	SPIKE-ah	Virgo	1.0
Pollux (30)	PAW-lux	Gemini	1.2
Fomalhaut (87)	FOAM-a-lot	Piscis Austrinus	1.2
Mimosa (89)	mim-OH-sah	Crux	1.2 - 1.3
Deneb (58)	DEN-eb	Cygnus	1.3
Regulus (42)	REGG-u-lus	Leo	1.4

Many a night from yonder ivied casement,
    ere I went to rest,
  Did I look on great Orion
      sloping slowly to the West.
Many a night I saw the Pleiades,
    rising thro' the mellow shade,
  Glitter like a swarm of fireflies
    tangled in a silver braid.
              — Alfred, Lord Tennyson

▶ Introducing the
Winter Sky

Stargazers willing to brave the chill of a clear winter's night will be rewarded with the most brilliant constellations of the year. The winter star map inside the front cover of this book will show you what's front and center, sky-wise.

You'll find the winter patterns in the south, and they're nearly all blockbusters. The seven constellations described in this chapter — Canis Major, Canis Minor, Gemini, Auriga, Taurus, Orion, and Lepus — are arranged in a circle. The Milky Way (if you can see it) cuts from northwest to southeast across the circle, grazing most of the groups but passing right through Auriga. West of the Milky Way is a bright blur of stars called the Pleiades, the most storied star cluster in the sky.

DAUGHTERS, DOVES, AND FIREFLIES

The glittering celestial fireflies evoked by Tennyson have appealed to storytellers for centuries. Ancient Greek skywatchers saw the Pleiades (*right*) as the seven daughters of the gods Atlas and Pleione. Atlas had run afoul of the gods, and as punishment was ordered to support the entire firmament on his shoulders. Naturally, the Seven Sisters were appalled. The king of the gods, Zeus, tried to appease the doting daughters by placing them in the heavens, but the girls remained grief-stricken at their father's cruel fate. If the Pleiades cluster appears blurry to you, it's because the sisters are forever weeping.

In American Native sky lore, a Kiowa legend records the plight of seven Indian maidens who, desperate to escape a pack of marauding bears, scaled a gigantic tree stump erected for their protection by the Great Spirit. The stump suddenly expanded upwards into the heavens, carrying the girls to safety. It is said that the tree stump still exists in the form of the famous 1,267-foot-high Devil's Tower in Wyoming. The butte's signature pattern of vertical striations shows where the big bad bruins clawed at the bark in their attempt to capture the terrified maidens.

Finding Your Way
in the Winter Sky

You'll find no better starting point for learning the winter sky than **Orion,** the most distinctive star pattern of *any* season. Orion is the only constellation that boasts two 0-magnitude stars. Another unique feature is a line of three 2nd-magnitude stars — the Belt of Orion — that bisects the Hunter's hourglass figure. Later in this section, you'll see how Orion can guide you to neighboring constellations. To start off, though, let this constellation be your key to a huge celestial signpost comprising three brilliant stars.

The first of these stars is 0.5-magnitude **Betelgeuse,** which marks Orion's northeastern shoulder. The second is −1.4-magnitude **Sirius,** the brightest star in the sky and the gorgeous *lucida* (brightest star) of **Canis Major** southeast of Orion. The third star is 0.4-magnitude **Procyon,** which dominates **Canis Minor** northeast of Canis Major. This stellar trio forms a prominent asterism called the **Winter Triangle.** If you extend the sides of the triangle in different directions, you can locate a variety of constellations.

▶ **A WORD ABOUT DIRECTIONS**

Celestial directions can be confusing. North doesn't always mean "up" and south doesn't necessarily mean "down." For example, on spring evenings the Big Dipper is above Polaris and on autumn evenings it is below Polaris — either way, the Dipper is always *south* of Polaris. So when I say a sky object is above or below (or left or right of) another sky object, I'm referring to when the objects in question are close to the meridian.

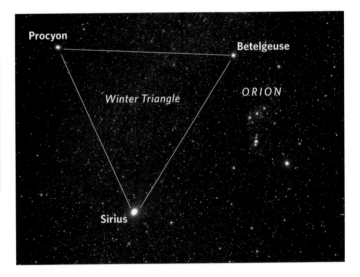

THOSE BRIGHT STARS

It seems obvious that the stars look brighter during winter because the sky seems so much clearer on frosty winter nights. While it's true that summer nights can be humid and hazy, the best summer nights are just as good as those of winter (or spring and fall). The reason the stars look brighter in winter is that a number of them *are* brighter. The winter constellations boast more zero- and first-magnitude stars (seven) than any other season.

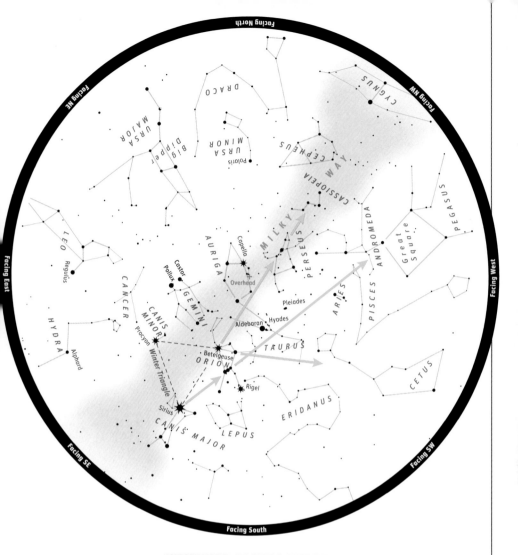

WESTWARD, BACK TO AUTUMN

Let's begin by following the Winter Triangle to a few autumn constellations before they depart for the season. A line westward from Procyon through Betelgeuse strikes the head of **Cetus,** a large constellation amid the receding autumn groups. Cetus is completely above the horizon at chart time, but sets within a couple of hours. The head of Cetus, marked by a hexagon of five stars, is the most recognizable part of the constellation.

A line northwestward from Sirius through Betelgeuse blazes a trail past **Auriga** overhead, then onward to **Perseus** and **Cassiopeia** in the northwestern sky. If you adjust the line from Sirius so that its trajectory aligns with the Belt of Orion, the line passes through Taurus (described in more detail on the next page) then onward in a west-northwest direction toward the tandem pair of **Andromeda** and **Pegasus.** The famous Great Square of Pegasus is angling toward the horizon but is still plainly visible. The Great Square is the "signpost" pattern for autumn; read more about its directional role on page 72.

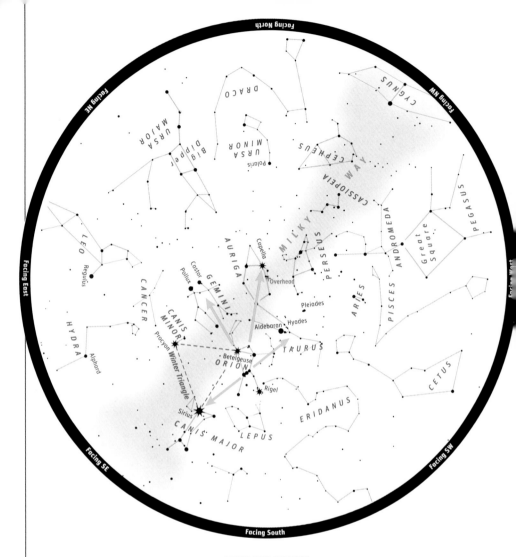

FOLLOW ORION

Cast your gaze toward the central part of the southern sky and the mighty Hunter, Orion. Sliding down Orion's Belt takes us in a southeastward direction toward **Canis Major** and brilliant **Sirius.** Aiming upwards along the Belt takes us in a northwestward direction toward **Taurus** and its bright red star, **Aldebaran.** An arrow northward from Orion's Belt aims toward **Auriga** and its prime star, **Capella.** A line northeastward from Rigel through Betelgeuse reaches the rectangular pattern of **Gemini** with its "twin" stars, **Castor** and **Pollux.**

And here's a "timely" trick to aid your recognition of the winter sky: Think of the eight best-known winter stars as the hours on a giant, oval clock face. Start with Capella. Almost directly overhead, Capella marks the 12 o'clock position. Aldebaran is at 2:00, Rigel marks 4:00, and Sirius is 6:00. Procyon chimes in at around 8:00, while Pollux and Castor together mark 10:00. The eighth star, Betelgeuse, is an off-center pivot point for the clock's hour and minute hands. You've never used the Winter Clock before? Well, it's about "time" you tried!

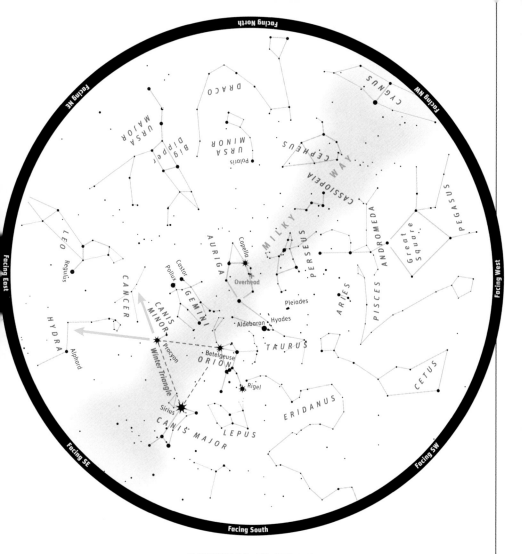

EASTWARD, TO SPRING

The Winter Triangle can give you a preview of the stars that will, in a few months, be better placed for viewing. A line northeastward from Sirius through Procyon reaches dim **Cancer** high in the east. You might not see Cancer from the city but at least you'll know where it's located. Rising below Cancer is **Leo**, whose bright stars — especially **Regulus** — should be plainly visible. Back to the Winter Triangle: A line towards the east-southeast from

Betelgeuse through Procyon passes by the head of **Hydra**, well up in the southeast. Below the head of Hydra, nearer the east-southeastern horizon, is **Alphard**, the only bright star in that region.

WINTER EXTRAS

Three minor constellations in the winter sky were not selected for descriptive treatment in this book. For the record, they are **Monoceros**, the Unicorn (a dim figure east of Orion); **Columba**, the Dove (a small constellation below Lepus); and **Camelopardalis**, the Camel (a vague pattern in the area north of Auriga and Perseus).

Orion, *the Hunter*

Big, bold, and bright, Orion has always dominated starlore, but in strikingly different ways. The ancient Egyptians considered him the soul of Osiris, the god of the afterworld. Egyptian wall reliefs show a stately Orion sailing through heaven in his celestial boat. By contrast, Greek mythology pictured the same star pattern as an aggressive warrior and hunter. When Orion boasted that he could kill every living creature, the Earth goddess Gaia arranged for a poisonous scorpion to engage him in battle. Orion got stung during the struggle and died. The two combatants were placed in the heavens 180° apart (Scorpius is a summer constellation; see page 62) to ensure that they would never tangle again.

As we saw on page 20, Orion cuts a grand figure in the sky. His northeastern shoulder is emblazoned by a member of the Winter

Triangle: **Alpha (α) Orionis,** or **Betelgeuse,** 430 light-years away. A red supergiant, Betelgeuse is roughly 800 times bigger than the Sun. The great star pulses irregularly between magnitude 0.4 and 1.3, though it is usually near the bright end of that range. His southwestern knee (or foot) gleams with 0.2-magnitude **Beta (β),** or **Rigel,** 770 light-years away. A blue supergiant, Rigel blazes with the power of 50,000 Suns and is one of the galaxy's most luminous stars. Orion's northwestern shoulder is marked by 1.6-magnitude **Gamma (γ),** or **Bellatrix,** while his southeastern knee is identified by 2nd-magnitude **Kappa (κ),** or **Saiph.**

Orion's midriff is aligned with the celestial equator. His famous belt is delineated by a row of 2nd-magnitude stars called **Alnitak, Alnilam,** and **Mintaka — Zeta (ζ), Epsilon (ε),** and **Delta (δ)** respectively. A line of fainter stars below his waist trace a gleaming sword (see next page), while a small triangle highlighted by 3rd-magnitude **Lambda (λ)** locates his head. Numerous dim stars, most of them hovering around 4th magnitude, outline a club and shield, which Orion raises in defence against Taurus, the Bull (see page 26).

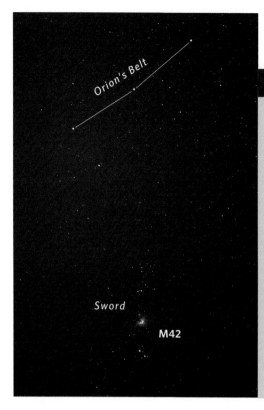

Orion's Belt

Sword

M42

In a dark sky, the Sword of Orion appears as a vertical row of three faint stars beneath the Hunter's belt. The fuzzy one in the middle is an *emission nebula* 1,500 light-years away called the **Great Orion Nebula,** or **Messier 42.** Binoculars turn M42 into a mist enveloping a pair of 5th-magnitude stars. The westernmost star is actually a tiny cluster (resolvable in telescopes) that causes the nebula to glow. South of M42 is 2.7-magnitude **Iota (ɩ) Orionis,** the Sword's brightest star and a fine double in scopes. Immediately southwest of Iota is a 4.7-magnitude star with a close 5.5-magnitude companion divisible in binoculars. North of M42 is a wide pair of 5th-magnitude stars, and just north of those two is a loose cluster easily seen in binoculars. There's a lot to explore in the Sword — don't miss it!

Lepus, *the Hare*

Pity little Lepus, the Hare: The timid creature hops above the horizon to check his surroundings and to his dismay discovers that a burly hunter (Orion) and his bloodhound (Canis Major) are uncomfortably close. But if the Hare is dangerously near these tormentors, he is happily free of another. An old tradition claims that rabbits detest the incessant screech of crows. When Lepus departs the sky each spring, his sensitive ears slip beneath the southwestern horizon just as Corvus, the Crow (page 51) rises in the southeast.

In Egyptian skylore, this constellation was connected with Orion. One interpretation holds that its main quadrilateral of stars symbolized the vessel in which the god Osiris (Orion) sailed down the Nile River. Another describes the pattern as a regal throne or "Chair of the Giant." Either way, Lepus is easily located directly beneath Orion. Four stars between magnitude 2.6 and 3.3 outline the quadrilateral. Also look for 3.6-magnitude **Gamma (γ) Leporis** a few degrees east-southeast of the quadrilateral. Just 29 light-years away, Gamma is a wide binocular double with a 6th-magnitude companion.

Taurus, *the Bull*

Greek legend links Taurus directly to Zeus. During one of his many amorous escapades, Zeus disguised himself as a pure white bull so he could cozy up to a princess named Europa. The bull nuzzled Europa on the seashore, then ferried her across the waves to Crete. There Zeus changed to human form and seduced the lass. A separate legend, also set in Crete, associates Taurus with the mighty Minotaur, a man-bull who was confined beneath the Minoan temple at Knossos. Alleged to be the bizarre offspring of King Minos himself, the monster was eventually slain by a Greek hero named Theseus (see page 47).

Star charts depict only the front half of the raging bull. His head and horns bent menacingly toward Orion, the animal appears ready to charge. The tips of his horns are marked by 3rd-magnitude **Zeta (ζ)** and 1.7-magnitude **Beta (β) Tauri,** or **Elnath.** First-magnitude **Alpha (α),** or **Aldebaran,** represents the animal's bloodshot eye. The rest of his face is outlined by a V-shaped asterism called the **Hyades.** According to legend, the Hyades were the

THE HYADES

The **Hyades** is an *open star cluster* — and a big one, too. Just 150 light-years away, the Hyades sprawl across 6° of sky. The V-shaped group appears to be dominated by fiery **Aldebaran** but it is 65 light-years distant and not physically part of the cluster. Aldebaran, an aging red giant some 360 times more luminous than our Sun, is a variable star whose brightness fluctuates slightly between magnitude 0.8 and 1.0. Compare Aldebaran with Betelgeuse, a star of similar color, brightness and variability in neighboring Orion (page 24). The brightest true member of the

Hyades is 3.4-magnitude **Theta²** **(θ²) Tauri,** which makes a naked-eye double with 3.8-magnitude **Theta¹ (θ¹)** — the two are marked as **Theta (θ)** on the chart. Your binoculars will sweep up several closer pairs, particularly along the south side of the V.

THE PLEIADES

The dipper-shaped **Pleiades,** or **Messier 45,** is the brightest open cluster in the sky. M45 is also known as the Seven Sisters because some sharp-eyed observers can count that many "sisters" without optical aid — but the cluster contains about 100 blue-white stars in all. Roughly three times the width of the full Moon, M45 is an ideal target for binoculars. To begin your exploration, look for a gentle arc of stars leading to 3rd-magnitude **Alcyone,** or Eta (η), the brightest member of the cluster. If you hold your binoculars steady, you can resolve a teensy triangle of stars immediately west of Alcyone. East of Alcyone, two stars symbolize the Pleiades' parents: 3.6-magnitude **Atlas** and **Pleione,** a variable star that ranges erratically between magnitude 4.8 and 5.5. (Try splitting this close pair without binoculars.) On the cluster's northwestern corner is an even tighter pair of stars with one name — **Sterope** — because to the eye they appear as a single 5.8-magnitude point of light. The Pleiades are 400 light-years away.

daughters of Atlas and Aethra. Entrusted by Zeus to nurse one of his sons to maturity, the Hyades were later rewarded with a position near their half-sisters, the **Pleiades,** an exceptional star cluster located in the Bull's shoulder. The Pleiades are described in more detail at right.

Taurus is a prominent member of the zodiac. The Sun reaches its northernmost point, the *summer solstice,* in eastern Taurus in late June. You'll spot planets here, too. When Mars wanders across Taurus, you'll find that its rusty-orange tint (and sometimes its brightness) is similar to that of Aldebaran. Finally, on an historical note, the English astronomer William Herschel discovered the Sun's sixth planet, Uranus, in eastern Taurus in 1781.

Auriga, *the Charioteer*

Two separate threads of classical mythology explain how a mere chariot driver became a major constellation. One is a complicated tale of deception and murder that I won't relate here. The other is a shorter and more pleasant story that celebrates human perseverance. It concerns a legendary character named Erichthonius, an early King of Athens. Disabled from birth, Erichthonius created the four-horse chariot as a means of getting around town. Soon lots of Athenians had these convenient conveyances. The King's inventiveness earned him a spot in the heavens as the star pattern we call Auriga.

However, the celestial version of Erichthonius seems more concerned with animal husbandry than chariot making. Old charts depict Auriga as a man holding a whip in one hand while balancing a small goat on his opposite arm. The animal is symbolized by 0.1-magnitude **Alpha (α) Aurigae,** or **Capella,** which means "little she-goat." Capella, the sparkling Goat Star, honours Almathea, who in one story was the Cretan goat who set aside her children to suckle the infant Zeus. A variation of this story suggests Capella represents a goat's horn that young Zeus broke off while playing. It was transferred to the heavens as *cornu copiae*, the Horn of Plenty. Fittingly, Capella clears the northeastern horizon soon after nightfall around harvest time.

Capella is 42 light-years away and second only to Sirius as win-

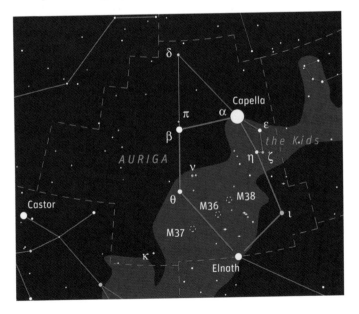

Auriga is easy to picture as a squashed pentagon, or stick-figure house, outlined by five prominent stars, including brilliant Capella. Elnath, a star that officially belongs to Taurus, marks the bottom of the house. The band of the Milky Way fills Auriga from basement to attic. Viewed with binoculars from a dark country location, the constellation is resplendent with powdery clusters of stars.

ter's brightest star. With a declination (or celestial latitude) of exactly 46°, Capella is the most northerly 0-magnitude star in the sky. In fact, if you live north of latitude 44° you can track Capella all year as a circumpolar star. Observers located on the 46th parallel will discover that Capella culminates directly overhead, at the zenith. Incidentally, Capella is a tight binary system featuring two yellow-white suns in close orbit around each other.

Two other stars in Auriga deserve special mention. The first is 1.7-magnitude **Elnath** (sometimes called Alnath), which seemingly establishes the southernmost point in the constellation. But you may recall from page 26 that Elnath marks the tip of one of the horns of the Bull. The star watchers of yesteryear considered Elnath as belonging to both Auriga and Taurus. Modern astronomers have conferred ownership of Elnath solely to Taurus, designating it **Beta (β) Tauri,** but most stargazers (including me) can't picture Auriga without it!

The second target, **Epsilon (ε) Aurigae,** is an unassuming 3rd-magnitude star that marks the northern end of a narrow triangle of stars known as the **Kids** located a few degrees southwest of Capella. Epsilon shines modestly only because it is 2,000 light-years away. This white-hot supergiant has an intrinsic luminosity of about 100,000 Suns, making it — like Rigel (page 24) and Deneb (page 58) — one of the most luminous stars in the galaxy. And that's not all. A mysterious companion star (or possibly a close pair of stars) orbits Epsilon once every 27 years. Astronomers think the companion might be shrouded in a cloud of gas and dust. As this dark agglomeration slowly passes in front of the brilliant primary star, we see Epsilon fade by eight tenths of a magnitude. The next "eclipse" of Epsilon Aurigae should begin in late 2009 and last for more than a year.

Gemini, *the Twins*

To ancient Greek stargazers, the rectangular outline of the constellation Gemini represented twin boys called **Castor** and **Pollux.** But the lads weren't true twins. According to legend, Castor was the mortal son of Queen Leda and King Tyndareus of Sparta, while Pollux, although the son of Leda, was sired by Zeus and thus immortal. Even so, the high-ranking half-brothers were dear friends. After Castor's death, Zeus reunited the boys in the night sky. Shining from their heavenly perch, Castor and Pollux reportedly guided the legendary Jason and his Argonauts in their search for the Golden Fleece (see page 82). Real-life mariners would turn to the twin stars as their guardians in bad weather. Gemini's brave spirit was symbolized as an electrical discharge seen in ships' rigging during storms. The phenomenon became known as Saint Elmo's fire.

The heads of Castor and Pollux are marked by **Alpha (α)** and **Beta (β) Geminorum,** respectively. Just 4.5° apart, these "twin" stars make a fine pair but, like their mythological namesakes, they aren't identical. Pure-white Castor, 52 light years distant, is magnitude 1.6. Yellowish-orange Pollux, 34 light years away, shines a little brighter at magnitude 1.2. The difference in color is obvious in binoculars. (Telescopes reveal an additional difference, for Castor is a superb binary star.) The rest of Gemini is a long rectangle formed by a half dozen stars ranging between 2nd and 4th magnitude. At the base of the rectangle, the Twins dip their 3rd- and 4th-magnitude "toes" into the Milky Way, which flows between Gemini and Orion. Using binoculars, can you spot the open cluster **Messier 35** just beyond **Eta (η)**?

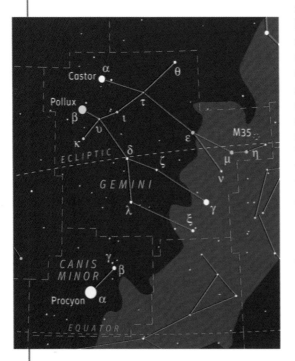

Gemini shares the northernmost sector of the zodiac with neighboring Taurus. Occasionally you'll spot a planet drifting inside Gemini, several degrees south of Castor and Pollux. Diminutive Pluto was discovered in Gemini in 1930 by the American astronomer Clyde Tombaugh.

Canis Minor, *the Little Dog*

As its name implies, Canis Minor, the Little Dog, is rather small. Ranked 71st in size among the 88 constellations, Canis Minor offers only two bright stars: 2.9-magnitude **Beta (β)** and 0.4-magnitude **Alpha (α) Canis Minoris,** better known as **Procyon.** The second-brightest member of the Winter Triangle (page 20), Procyon is just 11.4 light-years from Earth. The name Procyon means "before the dog," since it rises slightly ahead of Sirius, the true Dog Star in Canis Major, the Big Dog. Recognising that this starry duo were in the daytime sky together on the hottest days of the year, Egyptian skywatchers considered Procyon a co-host of the terrible dog days of summer (see next page).

Ancient Greek skywatchers pictured Canis Minor as a youthful companion for Canis Major, the full-grown bloodhound that belonged to Orion. Others saw Canis Minor as Maera, the faithful dog of Icarius, a wine maker who was brutally murdered. For leading Icarius' daughter, Erigone, to the scene of the crime, Maera was rewarded with a position in the sky. So were Icarius and Erigone. You can find out where they wound up by consulting page 48.

An Arabic story about Procyon faintly echoes a wonderful Chinese legend about celestial lovers separated by the summer Milky Way (page 53). In the Arabic tale, Procyon and Sirius were sisters living on the east side of the winter Milky Way. When Sirius eloped with her lover to the west, she left Procyon languishing on the wrong side of the river. There she remains to this day as Al Ghumaisa, the watery-eyed or Weeping One.

▶ WALKING THE DOG

It might be a stretch of the imagination but the next time you observe Gemini, pretend that the twin boys are on an evening stroll with their dog. The well-behaved dog is on a "leash" of six stars stretching from bright Castor in Gemini to brilliant Procyon in Canis Minor. The arc passes through 2.9-magnitude Beta (β) Canis Minoris, plus three fainter stars in Gemini: 3.8-magnitude **Iota (ι),** 3.5-magnitude **Delta (δ),** and 3.6-magnitude **Lambda (λ).**

Canis Major, *the Big Dog*

Canis Major, the obedient bloodhound of Orion, has a "nose" for notoriety. That's because its snout is marked by –1.4-magnitude **Alpha (α) Canis Majoris**, or **Sirius**. The brightest star in the night sky, pure-white Sirius has been the focus of lore and legend for more than 5,000 years. Egyptian skywatchers of the 3rd millennium BC based their calendar on the visibility of Sirius. Each year around the time of the summer solstice, the star's *heliacal* rising (near dawn after months of invisibility) announced the Egyptian New Year. More importantly, it heralded the annual Nile flood that irrigated farmers' fields. Because of this perceived relationship between the brightest star and the greatest river, Sirius was venerated as Sothis, the Nile Star.

But Sirius had a less savory reputation as well. It was the infamous **Dog Star,** an appellation in vogue long before the development of the constellation we know as Canis Major. The Egyptians believed that when the Dog Star was located in the summer daytime sky, it joined with the Sun to produce a period of extremely hot weather — the dreaded "dog days of summer" — that made animals thirsty and sick. Eventually, Sirius was regarded as the jackal-headed Anubis, lord of funeral rites and protector of tombs. It was also identified with the goddess, Isis, who accompanied Osiris, the god of the afterworld, symbolized by Orion.

The Greeks built upon these somber themes. To them, Sirius represented Cerberus, a three-headed, snake-ridden whelp assigned guard duty at the gates of Hades. The name Sirius, derived from a Greek word that meant "Scorching One," echoed the Egyptian belief that this brilliant star added unwanted heat during the summer. In Roman times, the Latin poet Virgil reviled the Dog Star as: " . . . that burning constellation, when he brings drought and diseases on sickly mortals, rises and saddens the sky with inauspicious light."

For modern stargazers, Sirius represents gleam, not gloom. The star is unmistakable both as the southern vertex of the Winter Triangle (page 20) and as the dominant star in Canis Major. However, Sirius

Patterns in the Sky

is brilliant mainly because it's only 8.6 light-years away. The surrounding constellation is outlined by a half-dozen much more distant stars. Chief among them are 1.5-magnitude **Epsilon (ε) Canis Majoris**, or **Adhara**; 1.8-magnitude **Delta (δ)**, or **Wezen**; 2nd-magnitude **Beta (β)**, or **Mirzam**; and 2.5-magnitude **Eta (η)**, or **Aludra.** These giant and supergiant stars are thousands of times more luminous than Sirius. Their apparent magnitudes are dimmed only by their distances. For example, Wezen and Aludra are 1,800 and 3,200 light-years away, respectively. If these stars were as close as Sirius, you'd need sunglasses to gaze at them!

The Milky Way spills toward the horizon just east of Canis Major. The 19th-century Finnish poet, Zakris Topelius, captured both the sweep of the Milky Way and the glory of Sirius when he penned a story about two lovers named Zulamith the Bold and Salami the Fair. The gods had Zulamith and Salami construct the glittering arch of the Milky Way. Working from opposite ends of the sky they toiled for a thousand years until, meeting in the middle, the ecstatic lovers. . .

> "*Straight rushed into each other's arms / And melted into one;*
> *So they became the brightest star / Great Sirius, the mighty sun.*"

The open cluster **Messier 41** is conveniently located 4° south of Sirius. M41 is a family of young stars like the Pleiades in Taurus (page 27), except that M41 is nearly six times farther away. Even at its distance of about 2,300 light-years, M41 is visible in a dark sky as a 4.5-magnitude patch of light about the size of the full Moon. Partly resolvable in steadily held binoculars, M41 contains several dozen blue-white stars plus a prominent, reddish-yellow pair near the middle. A 6th-magnitude star unrelated to the cluster shines on its southeastern periphery.

The Song of the Stars

We are the stars which sing,
  We sing with our light;
We are the birds of fire,
  We fly over the sky.
Our light is a voice,
  We make a road for spirits,
   For the spirits to pass over.
Among us are three hunters
  Who chase a bear;
There never was a time
  When they were not hunting.
We look down on the mountains.
  This is the Song of the Stars.

— Algonquin legend

▶ *Introducing the* *Spring Sky*

Mild spring nights invite the curious stargazer outdoors whenever the sky is clear. Like the other three seasonal charts, our spring star map inside the front cover of this book is a snapshot in time that will show you what's up. The 13 spring constellations described in this chapter lie across the south. The best ones are Leo, Virgo, Boötes, and, of course, Ursa Major. The Great Bear features prominently in the starlore of ancient Greece and remarkably, some early North American peoples also pictured that group of stars as a bear.

THE SPIRIT BEAR AWAKENS

A wonderful Micmac legend from Nova Scotia links the motion of Ursa Major to the cycles of nature. The Micmac bear is symbolized by the four stars outlining the bowl of the Big Dipper. Following behind him are seven hunter-birds, patiently stalking the beast around the sky. A robin leads the way. He is followed by a chickadee, a whiskey jack, a pigeon, a blue jay, and a pair of owls, each marked by its own star.

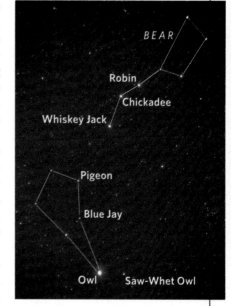

The hunt begins in spring, when bears emerge from hibernation. All through the spring and summer, the bruin ambles overhead. The hunters pursue it. In late summer the bear wanders lower in the sky, down by the northern treetops. One by one, the hunters begin dropping out of the hunt, as the seasonal motion sweeps them below the horizon. Finally, by autumn, only the robin, chickadee, and the whisky jack are left. They ambush the bear and the robin kills it with a single arrow. After the hunt, the bear's skeletal remains float magically through the sky. The bear's spirit senses the advance of winter, so it seeks out a hibernating bear and settles in. When spring arrives, the reborn Spirit Bear emerges and the hunt begins anew.

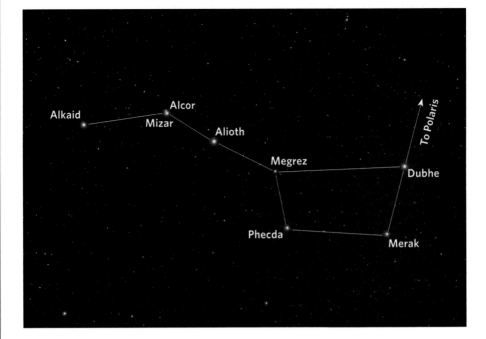

Finding Your Way
in the Spring Sky

As we saw on page 8, the **Big Dipper** can help you get oriented on any clear night of the year because two of its stars, called "the pointers," always aim at the North Star. As you become more familiar with the Big Dipper, you'll realize that it is another celestial signpost that can help you navigate the entire spring sky.

Of the seven stars outlining the Dipper, six of them are bright. The only exception, and it's a minor one, is 3.3-magnitude **Delta (δ) Ursae Majoris,** or **Megrez,** that joins the Dipper's handle to the bowl. If your sky is hazy, badly light polluted, or both, you might have difficulty spotting Megrez. The rest of the stars are easy. Outlining the bowl are 2.4-magnitude **Gamma (γ),** or **Phecda;** 2.4-magnitude **Beta (β),** or **Merak;** and 1.8-magnitude **Alpha (α),** or **Dubhe.** The crooked handle is delineated by 1.8-magnitude **Epsilon (ε),** or **Alioth;** 2.0-magnitude **Zeta (ζ),** or **Mizar;** and 1.9-magnitude **Eta (η),** or **Alkaid.**

These stars can help you locate not only the spring groups in the south, but also the last of the winter patterns departing in the west and a few summer ones rising in the east. So, are you ready to "dip" into the spring sky? The directions on the next three pages should be used at the same general time as the spring all-sky chart inside the book's front cover.

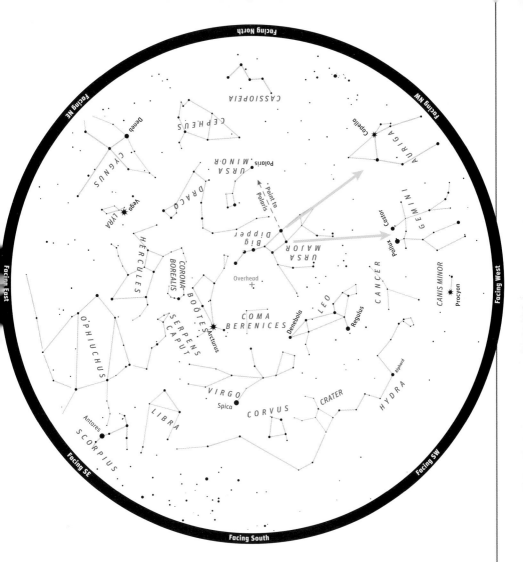

WESTWARD, BACK TO WINTER

Let's begin with the Big Dipper; it can direct us to a few lingering winter groups. A line across the top of the bowl from Megrez to Dubhe extended northwestward takes us to **Auriga** and its brilliant star, **Capella,** nearing the northwestern horizon at chart time. A line drawn diagonally across the bowl from Megrez through Merak, aims toward **Gemini** and its "twin" sparklers **Castor** and **Pollux** near the west-northwestern horizon. Hanging just above the western horizon, to the left of Gemini as you face west, is **Canis Minor** and its brilliant lucida, **Procyon.**

> ### ► BIG DIPPER, BIG CLUSTER
>
> The Big Dipper is so distinctive it doesn't seem possible that its seven stars can be aligned the way they are by chance. Well, it turns out that most of the Dipper stars are related. Five of them — Merak, Phecda, Megrez, Alioth, and Mizar — are between 78 and 84 light-years from Earth These five share a common speed and direction through space. Along with roughly 100 other stars all over the sky, the core members of the Dipper are part of the **Ursa Major Moving Cluster**, the nearest star cluster to Earth.

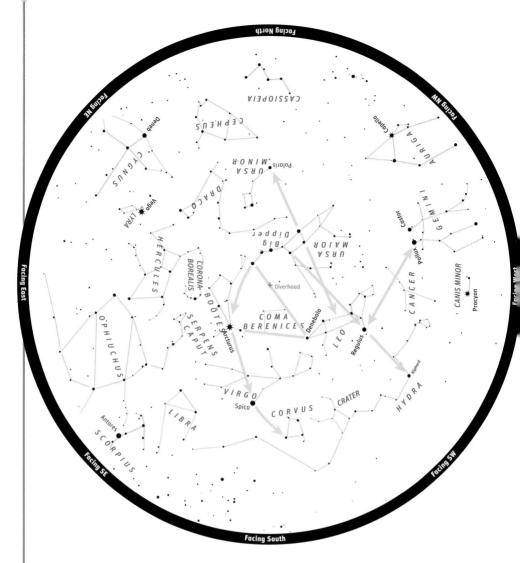

DIPPING INTO SPRING

The Big Dipper forms the bulk of **Ursa Major.** The pointer stars, Merak and Dubhe, aim northward toward **Polaris** in **Ursa Minor.** If we extend the pointers in the opposite direction, southward from Dubhe through Merak, we strike the back of **Leo.** A line from Megrez through Phecda aims close to **Regulus** in Leo. That line extended farther southwestward arrives at **Alphard,** in **Hydra,** not far above the southwestern horizon.

The Big Dipper produces a second route to the spring constellations. Its handle forms a shallow curve that arcs to **Arcturus,** the brilliant star in kite-shaped **Boötes.** The arc then speeds on to **Spica** in **Virgo.** Curling the line even more and extending the arc a bit further takes you to **Corvus** (fairly bright) and **Crater** (very faint). A triangle formed by Alkaid, Denebola, and Arcturus encloses the faint and formless **Coma Berenices.**

Finally, a line between Pollux in Gemini and Regulus in Leo passes through dim **Cancer.** The line passes just north of the spectacular **Beehive** star cluster.

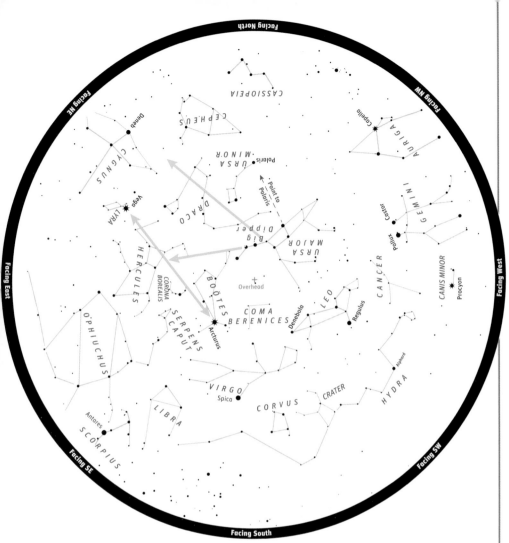

EASTWARD, TO SUMMER

The Big Dipper also points ahead to the stars of summer. For example, a line from Phecda through Megrez aims northeastward toward **Deneb** in **Cygnus,** also known as the **Northern Cross.** Just above Cygnus is tiny **Lyra,** with its brilliant star **Vega.** A line extended diagonally across the bowl from Merak through Megrez, then along the handle through Alioth and Mizar, aims roughly eastward toward **Hercules,** which is about halfway up the eastern sky. Finally, a line between Arcturus and Vega cuts across Hercules and **Corona Borealis,** a cute little "semi-summer" constellation that is described as part of our spring group.

SPRING EXTRAS

Three minor spring constellations were not selected for treatment in these pages. They are **Sextans,** the Sextant (a nondescript group between Leo and Hydra); **Leo Minor,** the Lesser Lion (a small, faint pattern between Leo and Ursa Major); and **Lynx,** the Lynx (a large but obscure constellation stretching across the void surrounded by Leo, Ursa Major, Gemini, and Auriga).

Ursa Major, *the Great Bear*

The third largest constellation in the heavens, Ursa Major patrols the circumpolar sky unchallenged. The Great Bear certainly overshadows the Little Bear nearby. Strangely, each bear has a long, bushy tail. A famous story in classical mythology tells why.

The goddess Hera suspected that her husband Zeus was having an affair with the beautiful maiden Callisto. Her hunch proved correct when Callisto bore Zeus a son named Arcas. Hera promptly transformed Callisto into a lumbering bear and banished her to the forest. Years later the orphaned Arcas, who had matured into a skilled archer, was hunting in the woods. Callisto recognized her beloved son and, forgetting she was a bear, rushed toward him. Arcas was about to fire an arrow at his own mother when Zeus intervened and changed the lad into a smaller bear. Zeus then grabbed the bruins by their tails and hurled them into the stars — stretching their tails in the process. In a final act of revenge, Hera ensured that the bears became circumpolar constellations so that they could never slip below the horizon and bathe in the cool waters of Earth.

The main part of the Great Bear — its body and tail — is outlined by the seven stars of the **Big Dipper**. If your sky is dark enough, you'll see another half-dozen stars of 3rd and 4th magnitude forming the bear's head and legs. Three nearly identical pairs of 3rd- and 4th-magnitude stars mark its huge paws.

ALCOR AND MIZAR

Stare at 2nd-magnitude **Mizar**, the middle star in the handle of the Big Dipper, and you might spot its wide companion, 4th-magnitude **Alcor**. Mizar is 78 light-years away and Alcor is three light-years further, so the stars aren't a true binary system — but they *look* like a pair. (Mizar itself is a spectacular binary in telescopes.) In the legendary bear hunt described on page 35, Alcor represents a pot in which to cook the catch. European skylore describes a wagon (the Dipper's bowl) pulled by three horses (the stars of the handle), with Mizar and Alcor as "the horse and rider." Early Middle Eastern writers claim that the eyesight of potential military recruits was tested on Mizar and Alcor. Test *your* eyesight on this terrific twosome.

Ursa Minor, *the Little Bear*

Ursa Minor's seven main stars are the same seven that form the Little Dipper. To help you determine its orientation in the sky, remember that the Little Dipper always curls toward the Big Dipper — as though the latter were pulling on the former. The Little Dipper's most famous member is 2nd-magnitude **Alpha (α) Ursae Minoris**, or **Polaris**, the star that pins the handle (and the Little Bear's long tail) to the north celestial pole. Because of its proximity to the pole, Ursa Minor has long been an aid to navigation. In the 3rd century BC, the Greek poet Aratus agreed that Ursa Major was "bright and easy to mark" but Ursa Minor was "better for sailors; for in a smaller orbit wheel all its stars."

After Polaris, the Little Dipper's next brightest stars — located at the end of the bowl — are 2nd-magnitude **Beta (β)**, or **Kochab**, and 3rd-magnitude **Gamma (γ)**, or **Pherkad**. Because this pair turns in a tight circle around Polaris, they're known as the Guardians of the Pole. The Little Dipper's other four stars, shining between 4th and 5th magnitude, usually aren't visible from the city (see page 15), though they appear readily in binoculars. Note that **Delta (δ)**, **Eta (η)**, **Zeta (ζ)**, and Pherkad are each accompanied by an unrelated fainter star. The 5th-magnitude neighbors of Zeta and Gamma glow pale orange in binoculars. Bright Kochab shines strongly reddish-orange — you might detect its warm hue without optical aid.

NOT THE NORTH STAR FOREVER

Polaris hasn't always been the North Star, nor will it remain so. The reason is that the Earth wobbles, or precesses, on its axis over a 26,000-year cycle. *Precession* will shift the north celestial pole closer to Polaris until the year 2100, then away from Polaris after that. When the Greek poet Aratus was writing about the constellations, the pole wasn't as close to Polaris as it is now. Three thousand years before that, the pole was near the star Thuban in Draco (see page 69). Precession also shifts the equinox and solstice points so that different constellations of the zodiac host these seasonal markers during different eras. For example, during the centuries when Thuban was the pole star, the summer solstice was in Leo (see next page).

Leo, *the Lion*

A formidable creature in classical mythology, the legendary lion was eventually slain by Hercules during the first of his Twelve Labours (see page 68). Leo's hide was considered impervious to stone and metal, so Hercules skinned the animal and wore the pelt as a suit of armour. In the sky, Leo's status as the celestial king of beasts is well earned, for the star pattern really does suggest a lion in regal repose.

The lion's head is outlined by a backwards question mark of six stars called the **Sickle.** The dot at the bottom of this asterism is 1.4-magnitude **Alpha (α) Leonis,** or **Regulus** ("Little King"), 78 light-years away. Blue-white Alpha is also known by its Roman title, **Cor Leonis,** which means "Heart of the Lion." The second-brightest star in the Sickle is 2.1-magnitude **Gamma (γ),** or **Algieba.** Binoculars reveal a 4.8-magnitude star just south of Algieba (which itself is a golden-yellow binary in telescopes). Likewise, 3.4-magnitude **Zeta (ζ)** is accompanied in binoculars by a 5.8-magnitude star underneath plus a slightly dimmer star above, making Zeta a wide triple star. The animal's hindquarters are represented by a triangle comprising 3.3-magnitude **Theta (θ),** 2.5-magnitude **Delta (δ),** and 2.1-magnitude **Beta (β),** or **Denebola** ("the lion's tail"). A 6th-magnitude star below Denebola shows in binoculars.

Leo is a prominent member of the zodiac. Planets can appear next to Regulus, and the Sun drifts southward through this region in late August. Due to precession, however, the situation was different in ancient times. Between 4,000 and 5,000 years ago, Leo was the most northerly of the zodiacal constellations and the Sun resided there around the time of the summer solstice. Believing that the

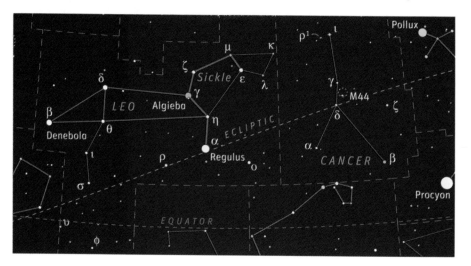

lofty lion imparted extra strength to the Sun god, Egyptian astrologers honoured Leo as the "house of the Sun." The annual flooding of the Nile River at the same time of year earned Leo even greater respect. No wonder the Egyptians sculpted huge statues of lions.

Cancer, *the Crab*

Cancer (a Latin word meaning crab) is hardly an appealing name for a constellation. It's tempting to dismiss this dim figure as an empty area between Gemini and Leo. But Cancer isn't empty, and it once enjoyed a degree of fame. Between 800 and 600 BC, the summer solstice was located in central Cancer. When it carried the Sun near the zenith on solstice day, the diminutive constellation was perceived as a blazing "Gate of Men" through which souls from heaven could descend to Earth and be reincarnated.

The starry crab resurfaced in classical mythology in a less elegant way. The goddess Hera had a grudge against mighty Hercules (to find out why, see page 68) and hatched a plot to disable him. Her secret weapon was the crab, but when the toothy crustacean tried to bite Hercules, he squashed it with one step. Nonetheless, Hera rewarded the crab with a place in the zodiac.

Cancer's basic pattern is an upside-down Y of five stars that range between magnitude 3.5 and 4.7. The northernmost star is **Iota (ι) Cancri,** a difficult and "unequal" binary with tightly spaced 4.0- and 6.6-magnitude components. Try splitting Iota with steadily mounted 10× binoculars. East of Iota is 6th-magnitude **Rho¹ (ρ¹),** a widely spaced pair. And any optical aid will reveal the Crab's finest prize: the famous Beehive star cluster.

A HIVE OF STARS

The **Beehive** star cluster, or **Messier 44,** is 520 light-years away. Although M44 is magnitude 3.1, the cluster's light is spread across more than 1° of sky so it looks fainter. In a dark sky, it appears as a grainy cloud. Binoculars will transform M44 into a celestial hive abuzz with starry bees, many of them shining between 6th and 7th magnitude. The Beehive is also called **Praesepe,** or the Manger. The Manger is flanked by 4.7-magnitude **Gamma (γ) Cancri,** or **Asellus Borealis** (Northern Donkey); and by 3.9-magnitude **Delta (δ) Cancri,** or **Asellus Australis** (Southern Donkey). The cluster between them makes an attractive pile of hay.

Canes Venatici,
the Hunting Dogs

In classical legend, Canes Venatici symbolized a pair of hunting dogs owned by Boötes, the Herdsman. Straining at their leashes, the eager bloodhounds led Boötes in his eternal pursuit of Ursa Major, the Great Bear, located immediately to the north. (For more about the Herdsman and his relationship to Ursa Major, see page 46.) Strangely, the more northerly of the two dogs, called Asterion, possesses no stars obvious to the eye — yet its name means "starry." The pattern symbolizing the southern dog is called Chara ("beloved") and is formed by two stars of unremarkable brightness. The fainter of the two is 4.3-magnitude **Beta (β) Canum Venaticorum,** or **Chara** (the name of the star and the dog). The brighter star is 2.9-magnitude **Alpha (α),** better known as **Cor Caroli,** a lonely light in the barren patch of celestial real estate bordered by Ursa Major, Boötes, Virgo, and Leo.

Cor Caroli is one of the few stars that honor a person who actually lived. Meaning "Heart of Charles," the name derives from the rather unwieldy title, *Cor Caroli Regis Martyris* in recognition of the martyred 17th-century English monarch, Charles I, who was executed in 1649. Charles' son, Charles II, restored the monarchy on May 29, 1660. An old legend claims that the starry Heart of Charles shone with extra luster on the eve of the new king's triumphant return to London. Perhaps the sky was especially clear that night, because any story that claims the star suddenly grew in brightness is an exaggeration — or an outright fabrication. In any case, the fancy four-part name was added to astronomical charts in 1673. Today, the star has been stripped of its *Regis Martyris,* but it's still full of heart.

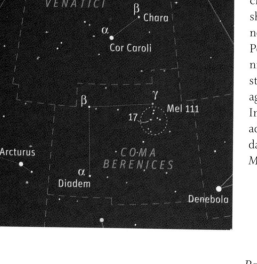

Coma Berenices,
Berenice's Hair

Coma Berenices is a formless scatter of stars, 4th magnitude and dimmer, between Leo, Virgo, and Boötes. The constellation name refers to Queen Berenice of Egypt. Unlike the many legendary characters immortalized in the night sky, Berenice was once a real person.

The wife of King Ptolemy III, Berenice lived in the third century BC. A legend about the queen states that she had long, blond hair. Concerned about her husband during a terrible war, she agreed to sacrifice her golden tresses to the goddess Aphrodite in exchange for his safety. When the King returned from battle, Berenice delivered on her promise and clipped her hair. Ptolemy demanded to see the golden gift in Aphrodite's temple but the hair had vanished! Later that night, the court astronomer, a clever fellow named Conon of Samos, drew the couple's attention to what's now called Melotte 111 — the delicate starfield east of Leo. Conon adroitly explained that Aphrodite had placed the shorn locks in the sky for all to see.

Sometimes this star pattern was pictured as a sheaf of wheat held by Virgo, who governed agricultural matters (see page 48). Alternatively, it was the tip of the tail of Leo, the Lion. Coma Berenices became a constellation in the 16th century when celestial cartographers deleted the lion's tail and awarded that sector of sky to the Egyptian Queen — or at least her hair. To help you remember it, think of **Diadem,** the name given to 4.3-magnitude **Alpha (α) Comae Berenices.** Diadem symbolizes the jewel-studded crown that Berenice wore before she sacrificed her lovely locks to the gods.

▶ **MELOTTE 111**

A spray of faint stars in Coma Berenices forms a loose open cluster called **Melotte 111.** Only 300 light-years away, Melotte 111 is 5° wide and is visible to the eye from a dark location. Binoculars will reveal two-dozen stars brighter than 8th magnitude. Among them is the wide double star, **17 Comae Berenices.** Its 5.3- and 6.6-magnitude components are an easy split in binoculars. On the cluster's northern periphery, 4.4-magnitude **Gamma (γ)** is actually a foreground star.

Boötes, *the Herdsman*

Like many constellations, Boötes is the subject of overlapping mythology. Sometimes he was pictured as an ox-driver tending his herd of oxen, which were symbolized by the stars of the Big Dipper. On other occasions, Boötes steered a plough (again, the Dipper), which he invented. The gods were so pleased with this advance in agriculture that they awarded the ploughman a spot in the heavens. A wider tradition saw Boötes slowly driving the Great Bear and Little Bear around the north celestial pole — a vital task that kept the heavens turning. Fittingly, the name of Boötes' brightest star, **Arcturus,** means "bear-keeper." Old atlases show the bear-keeper with two hounds (page 44). "What thinks Boötes," mused 19th-century Scottish philosopher, Thomas Carlyle, " . . . as he leads his Hunting Dogs over the zenith in their leash of sidereal fire?"

To modern stargazers, Boötes is a diamond-shaped kite flying in a spring breeze. The kite is outlined by six stars, five of them ranging

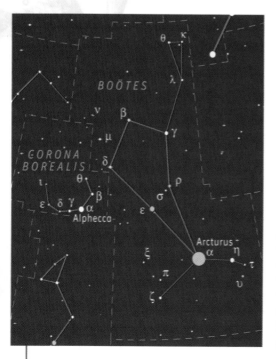

between magnitude 2.4 and 3.6. The sixth star is the constellation's 0-magnitude lucida, **Alpha (α) Bootis** — Arcturus — which marks the base of the kite. Three fainter stars west of Arcturus form the kite's tail.

Several treasures lie along the eastern side of the kite. The first is Arcturus itself. Just 37 light-years from Earth, Arcturus is the fourth-brightest star in the sky. Its deep-yellow hue is apparent to the eye and vivid in binoculars. Aim your binoculars at 3.5-magnitude **Delta (δ) Bootis,** a challenging double star 20° northeast of Arcturus. If you hold the glasses steadily, you might glimpse Delta's 8th-magnitude companion close by. A few degrees north-northeast of Delta is 4.3-magnitude **Mu (μ) Bootis** whose 6th-magnitude companion is easier to see. About the same distance farther along are orangey **Nu¹ (ν¹) Bootis** and white **Nu² (ν²) Bootis.** This widely spaced pair of 5th-magnitude stars is a cinch to separate.

Corona Borealis,
the Northern Crown

Corona Borealis is easy to miss. The little semicircle is formed by seven stars ranging from 2nd to 5th magnitude — most of them near the faint end of that range. Watching from your backyard without binoculars, you might see only the crown's brightest "gem," 2.2-magnitude **Alpha (α) Coronae Borealis** — also known as **Gemma** or **Alphecca**. But the petite pattern is sure to catch your eye from a country observing site.

According to legend, the crown belonged to the wine god, Bacchus, and (momentarily) to a lovely Princess named Ariadne. Ariadne was the daughter of King Minos of Crete, a vengeful man who fought a series of wars against Greece. Minos captured Athenian youths to feed to his Minotaur, the ravenous half-man, half-bull that roamed an escape-proof labyrinth in the King's huge cellar.

Enter Theseus, heir to the Athenian throne. Brave Theseus insisted he could navigate the Minos maze and slay the hideous creature within. Ariadne fell in love with Theseus and aided his cause by loaning him a ball of golden yarn. Theseus marked his path through the labyrinth by trailing the glowing thread behind him. After destroying the Minotaur, he retraced his steps, collected Ariadne, and set sail for the voyage back to Greece. The couple stayed overnight on the island of Naxos, which happened to be Bacchus' personal estate. Alas, the next morning Theseus was forced to carry on alone, tricked by Bacchus, who desired Ariadne for himself. The scheming deity won her hand by offering his jewel-studded crown as a wedding gift. Later, Bacchus tossed the crown into the heavens in honour of his beautiful bride.

I prefer a different version of the story in which the princess was wearing her own royal crown when she met Theseus. Ariadne offered our hero the dazzling tiara as a kind of headlight for his treacherous journey through the maze. The mission completed, Theseus and Ariadne got married and lived happily ever after. Today we see her crown among the stars, a fitting tribute to the loving lady of ancient Crete.

Virgo, *the Maiden*

Straddling the celestial equator, Virgo is the sky's second-largest constellation and biggest member of the zodiac. Classical mythology awarded Virgo two powerful portfolios: agriculture and justice. In the first, Virgo was Demeter, the corn and grain goddess. She would appear each spring when the crops were planted, remain in view throughout the growing season, and then sink into the sunset around harvest time. The Sun would then pass through Virgo and cross the *autumn equinox*. During ancient Greek times, the equinox was located about 10° east of brilliant **Alpha (α) Virginis,** or **Spica,** which means "ear of wheat."

Virgo was later identified with Astraea, the goddess of justice, a link apparent in her Roman name, Justa. Some old star charts portrayed a blindfolded Astraea holding the scales of justice, which were symbolized in neighboring Libra (see next page). Astraea used the sacred balance to assess people who had recently died and thus determine whether their souls should rise to heaven or descend to Hades.

A separate legend portrays Virgo as Erigone, the loving daughter of Icarius, a pioneering Greek winemaker who learned his trade from the wine god, Bacchus. Icarius was murdered by his neighbors after they fell sick from drinking his powerful nectar. When the family dog led Erigone to the grim remains, she died of

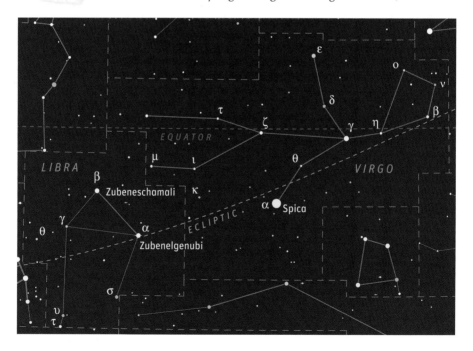

grief on the spot. Bacchus placed Erigone in the heavens as Virgo, while Icarius became Boötes just north of Virgo. The dog was rewarded, too (see page 31).

Expansive Virgo displays no identifiable pattern to city stargazers. Most of its dozen main stars shine between 3rd and 4th magnitude. The exception is 1st-magnitude Spica, a blue-white binary star 260 light-years away. Planets occasionally pass close to Spica on their long journey across the constellation. The autumn equinox, which has been precessing westward from the Libra-Virgo border since 700 BC, is *still* in Virgo.

Libra, *the Scales*

Look carefully if you want to spot Libra low in the southern sky. Located between sprawling Virgo and glittering Scorpius, this modest member of the zodiac is small in size and low in luminosity. In one sense, however, Libra is special. While the other eleven star patterns in the famed "ring of animals" represent various creatures and people, Libra is the sole sign that symbolizes an inanimate object; namely, the Scales. But this is only because the constellation went through a change of identity.

Originally, these stars were regarded as the claws of the Scorpion, the poisonous creature immediately east of Libra. Between 3,000 and 4,000 years ago, the Sun entered what was then considered western Scorpius around the time of the autumn equinox. As the Sun plunged southward across the celestial equator, day and night were of equal length. This important event motivated skywatchers to cut off the pincers and, in their stead, create a heavenly weigh-scale that symbolized the temporary balancing of day and night. Libra's role later expanded to become the Scales of Justice held by Virgo (see opposite).

Libra is a boxy pattern comprising half a dozen 3rd- and 4th-magnitude stars. The constellation's two brightest markers, 2.8-magnitude **Alpha (α)** and 2.7-magnitude **Beta (β) Librae,** certainly aren't prominent but are notable for other reasons, not the least of which is their delightful, tongue-twisting names. Alpha Librae is **Zubenelgenubi** (zoo-ben-el-jen-NOO-be) and Beta Librae is **Zubeneschamali** (zoo-ben-es-sha-MAY-lee). These colorful monikers derive from the Arabic for "southern claw" and "northern claw," a legacy from the days when the stars of Libra represented the front end of Scorpius. If you aim your binoculars at Zubenelgenubi, you'll discover that it's accompanied by a 5.2-magnitude companion, nicely separated. Take a look at Zubeneschamali as well. Most observers agree that it appears somewhat greenish, a rare hue among naked-eye stars.

Hydra,
the Water Snake

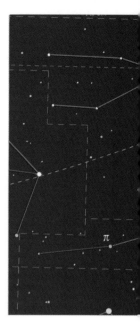

To city skywatchers, Hydra, the Water Snake, is challenging to observe. Yet this slender star pattern is the largest of the 88 constellations, stretching more than one quarter of the way around the celestial sphere. And Hydra is one scary serpent, too. According to legend, Hydra was a man-eating snake that patrolled the marshes of ancient Greece. The creature had nine heads. If one was cut off, two others would grow in its place. Yet Hydra lost all nine noggins when the strongman Hercules (described on page 68) beheaded the beast during the second of his Twelve Labours.

Hydra's father, Typhon, had even more heads and reportedly stood as high as the mountains. Greek storytellers occasionally blended parent and child into the same constellation, perhaps because they saw Hydra rise almost vertically out of the horizon, its head(s) reaching for the zenith. This may have inspired a story in which Typhon launched an attack on Mt. Olympus. Zeus struck back with his thunderbolts, finally subduing Typhon by placing gigantic Mount Etna on top of him. Even today, the famous volcano in Sicily trembles from the writhing prisoner inside.

The constellation Hydra depicts a *female* water snake. She has a male counterpart, called Hydrus, who swims in the deep southern hemisphere. Most of Hydra is submerged in the south as well. Only her head sticks above the celestial equator, as though she were coming up for air. Outlined by a small formation of 3rd- and 4th-magnitude stars, the head is located just south of Cancer. From there, Hydra's body stretches southeastward under Leo and Virgo before terminating near the southwestern corner of Libra. All except one of her stars shine between 3rd and 5th magnitude. The loner is 2nd-magnitude **Alpha (α) Hydrae,** or **Alphard,** which means "solitary one in the serpent." The name is apt, for no other star in the region rivals Alphard. Reddish-orange Alpha is also **Cor Hydrae,** the reptile's glowing heart.

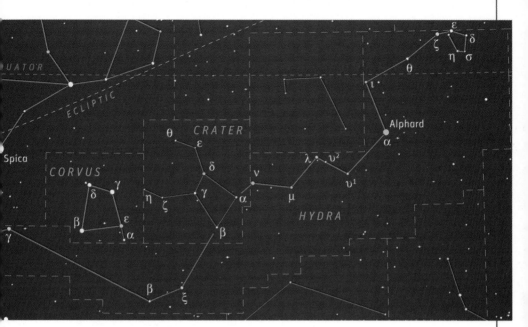

Corvus, *the Crow*
& Crater, *the Cup*

Corvus, the Crow, and Crater, the Cup, are balanced on the coils of Hydra, the Water Snake. The Roman poet, Ovid, linked all three constellations in a legend, which asserts that crows were once pure-white and possessed a beautiful singing voice.

According to the story, the Greek Sun god, Apollo, dispatched his servant, the crow, with a sacred chalice to fetch water. The bird got distracted by a ripening fig tree and returned quite late. The crow lied about his tardiness by claiming he was attacked by a water snake, whose remains he held in his claws. Apollo didn't believe him. Bent on punishment, Apollo reduced the bird's tuneful voice to an awful screech and changed its plumage from white to black. A different legend asserts that the loyal crow told Apollo his wife was being unfaithful to him. Apollo went into a rage and turned the bird black as coal. Talk about shooting the messenger!

Apollo wasn't done yet. He cast a spell of unquenchable thirst on Corvus then flung all three items — bird, cup, and reptile — into the sky. Corvus alighted on Hydra's back just east of the chalice, which was brimming with water. Alas, the sky's westward drift keeps the prize forever out of the crow's reach.

Corvus' boxy pattern, a quadrilateral of 3rd-magnitude stars, is not at all birdlike (crow or otherwise) but is at least visible in a city sky. By contrast, Crater's 4th- and 5th-magnitude stars outline an ornate celestial goblet that urban observers can't see at all.

Silently I watch the River of Stars
Turning in the Jade Vault ...
Tonight I must enjoy life to the full,
For if I do not,
Next month, next year,
Who can know where I shall be?
— Su T'ung-Po

▶ Introducing the Summer Sky

Nothing soothes the soul quite like gazing up at the stars on a warm summer's evening. The seasonal star map inside the back cover of this book will set the summer sky scene for you.

The 11 summer constellations described in this chapter fill the south and include large but inconspicuous patterns, such as Draco, Hercules, Ophiuchus, and Serpens, plus bright ones such as Cygnus, Aquila, Sagittarius, and Scorpius. The latter group also hosts the band of the Milky Way. Subtle, delicate, and mysterious, the Milky Way has been stirring people's imaginations for thousands of years.

THE CELESTIAL RIVER

According to an ancient Chinese legend, the Milky Way was known as the Celestial River. The legend also records that the estate of the Chinese Sun god sprawled along one bank of the river. One day the Sun god's daughter, the Weaving Princess, spotted her father's handsome cow-herder tending his cattle by the riverbank. She fell in love with him, and despite his low station they were married.

Unfortunately, the lovers became so preoccupied with each other they neglected their royal duties. After repeated warnings, the Sun god banished the groom to one side of the Celestial River and ordered his daughter back to her loom on the other. In time, however, the gods allowed them to meet. They decreed that on the seventh day of the seventh month each year, all the magpies in China should flock to the Celestial River and make a bridge so the estranged couple could cross. Alas, by dawn the pair would be ordered apart. The morning dew is the lovers' tears of sorrow at the prospect of separation for yet another year.

The main players in this bittersweet love story are symbolized by Vega (the Weaving Princess) and Altair (the cow-herder), two brilliant stars on opposite sides of the Milky Way. A fainter star called Albireo, located in the band of the Milky Way between them, symbolises the first magpie arriving to form the bridge. All three stars are high in the south around midnight on July 7.

Finding Your Way
in the Summer Sky

▶ BRIGHTNESS VS.
DISTANCE

The three stars of the Summer Triangle demonstrate that a star's brightness is not necessarily an indication of its distance. Zero-magnitude Vega is 25 light-years from Earth. Altair appears somewhat dimmer than Vega, yet it is only 16.8 light-years away. But Deneb is the shocker. Deneb is only half a magnitude fainter than Altair, yet it is 1,600 light-years away. A blue supergiant, Deneb's luminosity is roughly 60,000 times that of the Sun!

Your starting point for navigating the summer sky is brilliant **Vega,** which shines near the zenith on summer evenings. (For observers along the 39th parallel, Vega actually crosses the zenith.) The Earth itself seems to be homing in on this gorgeous sapphire. The Sun's motion around the Milky Way Galaxy is carrying our planet in the general direction of Vega at a velocity of 20 kilometers per second (about 12 miles per second) relative to the nearby stars. Also, because the Earth wobbles on its axis over a period of 26,000 years, the north celestial pole is slowly shifting away from Polaris, towards Vega. Vega will be the North Star around 12,000 years from now, though it won't be as close to the celestial pole as Polaris is today.

The fifth brightest star in the heavens, 0-magnitude Vega is the dominant sparkler in tiny **Lyra.** East of Lyra is 1.3-magnitude **Deneb,** the brightest star in **Cygnus.** South of Cygnus is 0.8-magnitude **Altair,** the leading star in **Aquila.** This trio of stars forms a huge asterism called the **Summer Triangle.** By extending the imaginary lines that connect these three stars, the Summer Triangle becomes yet another valuable "signpost" that points to constellations all over the summer sky — plus a few belonging to spring and autumn — as you will see on the following three pages. The directions given are designed for the same evening times listed on the summer all-sky chart inside the back cover.

Patterns in the Sky

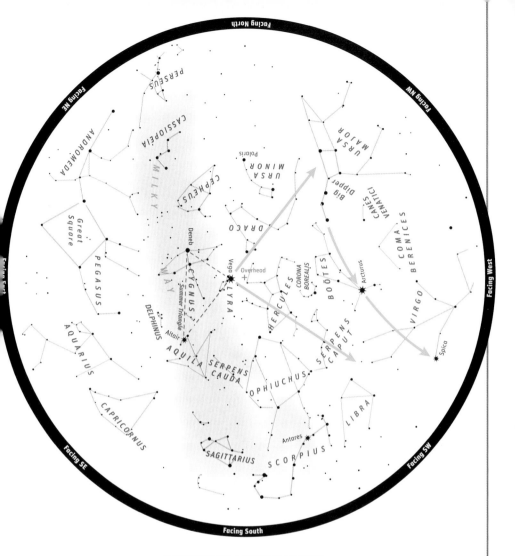

WESTWARD, BACK TO SPRING

Let's start by allowing the Summer Triangle to point us to a few spring constellations that are now departing for the season. A line from Deneb through Vega aims toward the southwest where two zodiacal constellations, dim **Libra** and brighter **Virgo,** hover just above the horizon. These two star groupings are often hard to find after mid-summer, or more than an hour or two after dusk in early summer, because they are too low in the sky and can be completely obscured by horizon haze. Get them while you can.

Back to the Summer Triangle: A line from Altair through Vega slices northwestward across **Draco** (grazing the dragon's head) all the way to the **Big Dipper.** Recall from our spring signpost directions (pages 36-39) that by following the Dipper's handle, you can "arc to **Arcturus**" at the base of kite-shaped **Boötes** halfway down the western sky. If you want to "speed on to **Spica**" in Virgo you'd better hurry because it's sinking fast.

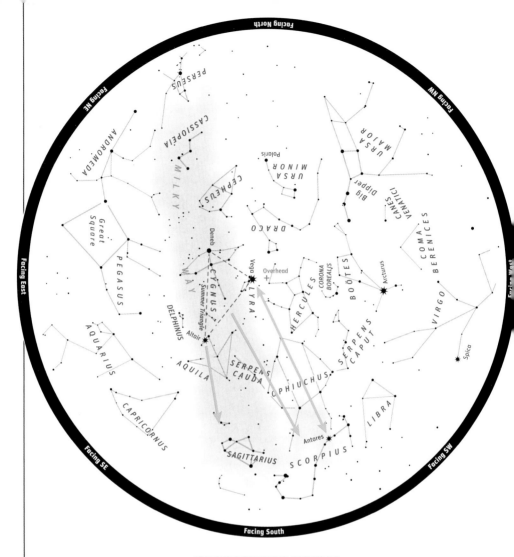

FLYING SOUTH TO SUMMER

Return again to the Summer Triangle. Following a line from Deneb, through Altair, and into the south will get you to **Sagittarius.** Deneb is part of Cygnus, the celestial Swan (see page 58), and if you follow the bird's direction of flight toward the southern horizon, you'll "fly" past Aquila and into **Scorpius.** The 1st-magnitude star **Antares** in Scorpius should stand out above all others in the region. A line between Vega (in the Summer Triangle) and Antares passes through the sprawling — though dim — tandem of **Ophiuchus** and **Serpens.**

If you can get away to a country observing site sometime during the summer, you'll find to your delight that Cygnus and Aquila both have glittering flight paths — the Milky Way. The band of the Milky Way is a signpost in its own right. It arches right across the sky, connecting a string of constellations from **Cassiopeia** and **Cepheus** in the northeast, through Cygnus and Aquila high in the south, to Sagittarius and Scorpius near the southern horizon.

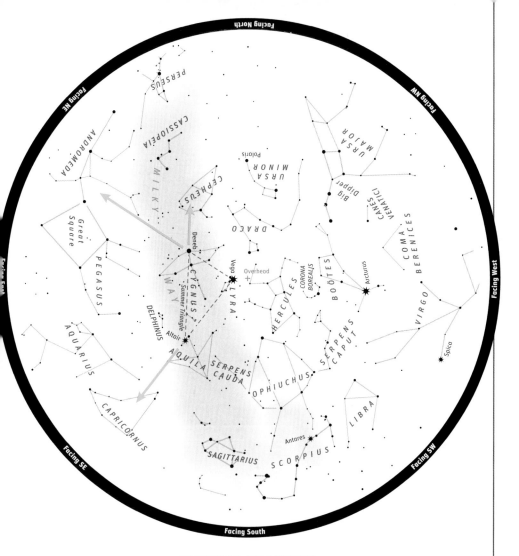

EASTWARD, TO AUTUMN

The Summer Triangle can help us look ahead toward the autumn constellations. For example, a line from Vega through Altair aims southeastward to **Capricornus,** an indistinct member of the zodiac near the southeastern horizon. A line from Altair through Deneb aims northward to the faint autumn pattern of **Cepheus.** His neighbour, **Cassiopeia,** lies half way between Cepheus and the northeastern horizon. Finally, a line extended from Vega through Deneb points northeastward toward **Andromeda** and (connected to it) **Pegasus.** A glance at the summer all-sky chart will confirm that the Great Square of Pegasus is just clearing the northeastern horizon around nightfall.

SUMMER EXTRAS

Three minor constellations in the summer sky do not appear in this section. They are **Vulpecula,** the Fox (a faint pattern between Sagitta and Cygnus); **Scutum,** the Shield (a tiny, formless group between Aquila and Sagittarius); and **Corona Australis,** the Southern Crown (another small constellation beneath Sagittarius).

Cygnus, *the Swan*

The story of Cygnus, the Swan, begins with two brothers named Cycnus and Phaethon, who were sons of Apollo, the Sun god. One day, Phaethon borrowed the fiery chariot in which Apollo rode from sunrise to sunset. In no time, the reckless youth managed to scorch both heaven and earth. The gods were not amused. Zeus corrected the chariot's trajectory, then fried Phaethon with a thunderbolt and dropped him into a river. Horrified, Cycnus plunged into the water to retrieve his brother's charred remains. In compassion, the gods immortalized brave Cycnus as the graceful constellation Cygnus in the summer Milky Way. Fittingly, the band of the Milky Way is said to be the trail blazed by Phaethon during his wild ride across the heavens.

The Swan flies southward. Its head, at the bottom of the constellation, is marked by 3rd-magnitude **Beta (β) Cygni,** better known as **Albireo.** Albireo is a fine double star with yellow and blue components that can be resolved in large (7 × 50 or 10 × 50) binoculars. The Swan's tail is marked by a member of the Summer Triangle — blue-white, 1.3-magnitude **Alpha (α),** or **Deneb.** The wings of

LUMINOUS SWAN: A bright portion of the Milky Way runs the length of Cygnus. On the right is the parallelogram of Lyra, described on the next page.

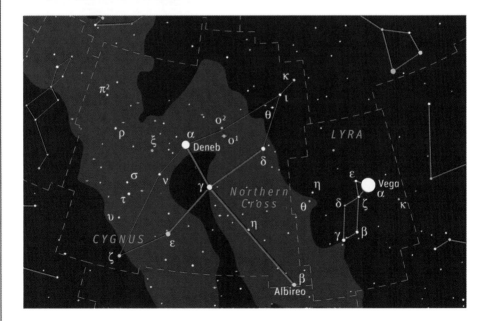

Patterns in the Sky

the Swan are outlined by 2.5-magnitude **Epsilon (ε)**, 2.2-magnitude **Gamma (γ)**, and 2.9-magnitude **Delta (δ)**. These stars also join with Deneb, Albireo, and fainter **Eta (η)** to form a prominent asterism called the **Northern Cross.**

If you aim your binoculars 5° west of Deneb (in the Swan's wing) you'll encounter a pair of 4th-magnitude stars called **Omicron¹ (o¹)** and **Omicron² (o²)** Cygni. Orangey-red Omicron¹ has two blue companions. The first is a 5th-magnitude star immediately northwest, and the other is a 7th-magnitude star very close by, to the south. This colourful triplet, plus Omicron², will fit easily in your binoculars' field of view.

Lyra, *the Lyre*

Lyra was the lyre, or harp, played by Orpheus, a gifted musician whose sublime compositions charmed the Olympian deities. When Orpheus died, the gods dispatched a vulture to retrieve the cherished harp — ostensibly to display it in the heavens. But some say that the Olympians took the lyre for themselves and placed the vulture in the sky. Brilliant blue-white **Vega** reflects this ambiguity: While it is often called the Harp Star, the name Vega has Arabic roots that translate as "swooping vulture" or eagle. Some old charts picture the constellation as a bird-of-prey clutching the lyre.

As the brightest member of the Summer Triangle, 0-magnitude **Alpha (α) Lyrae**, or Vega, draws our attention to Lyra's petite parallelogram of 3rd- and 4th-magnitude stars. If you have binoculars, you'll notice that the corners are busy places. The northeastern corner is marked **Delta (δ)**, a colorful double consisting of a bluish, 5.6-magnitude star and its reddish, 4.2-magnitude companion. A bigger challenge is 4.3-magnitude **Zeta (ζ)**, which marks the corner nearest Vega. Zeta harbors a close 5.7-magnitude companion. At the Harp's southeastern corner, 3.3-magnitude **Gamma (γ)** is accompanied by a dimmer star under it. **Beta (β)** on the southwestern corner is attended by two dim stars below. Keep your eye on Beta, for it is an *eclipsing binary* whose magnitude varies continuously over a period of 13 days. When Beta is at maximum light it's as bright as neighboring Gamma; at minimum, it's one full magnitude dimmer.

Lyra's most famous double, **Epsilon Lyrae,** appears to the eye as a 4th-magnitude star less than 2° northeast of Vega. However, binoculars will split Epsilon into 5.0-magnitude **Epsilon¹ (ε¹)** and 5.3-magnitude **Epsilon² (ε²)**. Telescopes reveal that ε¹ and ε² are themselves tightly twinned, so Epsilon Lyrae is known as the "Double-Double."

Aquila, *the Eagle*

Classical mythology is replete with tales of Aquila's service to the Olympian gods. In one instance, Zeus had the Eagle neutralize Aesculapius, the god of medicine (the reason why can be found on page 66). In another, the royal raptor was ordered to find Prometheus and gnaw out his liver. (Prometheus had defied Olympian decree by introducing fire to humankind.) Aquila also recruited servants to work for the gods. One day, the giant bird swooped down to Earth, abducted an innocent shepherd named Ganymede, and carried him back up to Mount Olympus (the boy's fate is revealed on page 86).

The constellation's brightest star is pure white, 0.8-magnitude **Alpha (α) Aquilae**, or **Altair** ("flying eagle"). A member of the Summer Triangle, Altair marks the gleaming eye of the Eagle, while two dim flanking stars complete the Eagle's head. The chart shows several more 3rd- and 4th-magnitude stars that outline the bird's broad wings and even a short tail. Like its avian friend, Cygnus, Aquila's main pattern is easily traced in a suburban sky.

DIRTY MILK: The Milky Way flows through all of Sagitta and most of Aquila. Note the Great Rift, an irregular dark lane of interstellar dust that runs the full length of the Milky Way in this image. The bright patch at bottom-right is the Scutum Star Cloud in the dim constellation of Scutum, the Shield.

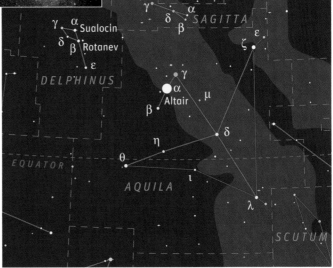

Patterns in the Sky

Delphinus, *the Dolphin*

To the sea-going Greeks of ancient times, Delphinus was the Sacred Fish. Several legends portray the Dolphin as a good-natured creature capable of selfless heroism. In one instance, Delphinus rescued the maiden Amphitrite, who was lost at sea and presumed drowned. Poseidon, the King of the Sea (and Amphitrite's suitor), expressed his gratitude by creating a likeness of the mammal in the heavens. He did a good job: Delphinus is a compact, diamond-shaped pattern that really does suggest a dolphin or porpoise leaping over the waves.

The five stars forming the Dolphin's body and tail all shine around 4th magnitude, so you might need binoculars to confirm your sighting. The two brightest stars — 3.8-magnitude **Alpha (α)** and 3.6-magnitude **Beta (β) Delphini** — are named **Sualocin** and **Rotanev** respectively. These cryptic monikers first appeared in an Italian star catalog in 1814. Their meaning stumped astronomers for years until someone reversed the letters and saw "Nicolaus Venator," the Latinised name of one Niccolò Cacciatore who, in 1814, was an assistant at the Palermo Observatory in Sicily. Cacciatore became the only person to name two stars after himself — and get away with it!

Sagitta, *the Arrow*

Sagitta ranks 80th in size among the 88 constellations but it figures in many classical legends. According to one story, the Arrow belonged to mighty Hercules. During the fifth of his Twelve Labours (see page 68), Hercules fired the projectile at a trio of monstrous, man-eating birds. Each one is a constellation near Sagitta. Aquila, the Eagle, soars beneath the eastward-aiming Arrow and Cygnus, the Swan, flies north of it. The third bird is Lyra, usually identified as a harp but once considered a vulture (page 59). Sagitta is even smaller than Delphinus but roughly as bright and fairly easy to identify. Its half-dozen main stars will fit in a single binocular field. Binoculars will show that the Arrow is lodged in the Milky Way.

Scorpius, *the Scorpion*

The Scorpion had been assigned by the Earth goddess, Gaia, to zap the unruly hunter Orion (see page 24). After a successful "sting" operation, the creature's reward was a permanent home in the summer Milky Way — opposite from Orion so that the two adversaries could never meet again.

Slender Scorpius is arguably the finest star pattern in the summer sky. A north-south line of 2nd- and 3rd-magnitude stars forms the Scorpion's head. **Alpha (α) Scorpii,** or **Antares,** whose brightness wavers between magnitude 0.9 and 1.2, pinpoints the creature's heart. About 430 light-years away, Antares is a red supergiant some 500 times the size of the Sun and a whopping 12,000 times its luminosity. Below Antares, a graceful curve of bright stars shapes the Scorpion's upraised tail (though it might not clear your horizon; see page 9). A double-strength "stinger" is provided by 1.6-magnitude **Lambda (λ),** or **Shaula,** and 2.7-magnitude **Upsilon (υ),** or **Lesath.** These stars are not related; Shaula is roughly 700 light-years away while Lesath is nearly 200 light-years closer.

The chart shows three doubles that will test your eyesight. In the head of Scorpius is a wide pair, 4th-magnitude **Omega (ω).** However, you might need binoculars to separate the tighter components of 3rd-magnitude **Mu (μ)** in the Scorpion's back. Lower still, and similarly tight, is lovely **Zeta (ζ),** which features a 3.6-magnitude orange star and a 4.7-magnitude bluish star. The Zeta pair combines with a 6th-magnitude star to form the "head" of the **False Comet.** Two star clusters — one small and dense, the other large and loose — spread northeastward from Zeta to suggest the comet's curving tail. The clusters look superb in binoculars, though they appear low in the sky. The declination of Zeta is −42°, which means that if you live north of latitude 48°, the star remains below your horizon.

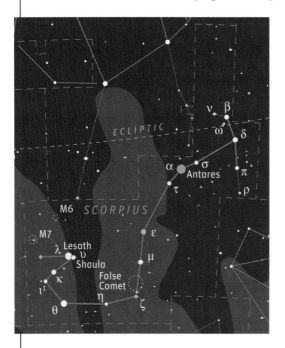

Two other aspects of Scorpius deserve mention. First, planets occasionally track across the northern half of the constellation because Scorpius is a member of the zodiac. Sometimes a planet drifts close to Antares. The name Antares means "rival of Mars," a reminder from ancient times that a conjunction between the red planet and the red star at-

DARK SKY SCORPION: The slender "fishhook" of Scorpius is buried in the Milky Way. Note the clusters M6 and M7 left of the Scorpion's tail, plus the False Comet in the tail itself. The brilliant object to the left of Antares is Mars. Antares is no rival of Mars on this occasion!

tracted attention. Second, if you observe away from the city, you'll notice the Milky Way traversing the southern half of Scorpius. As we saw on the previous page, this region is rich with star clusters. Two more are described below.

STINGING STAR CLUSTERS

Two fine open clusters shine near the tail of Scorpius. **Messier 7**, the larger of the two clusters, is a 3rd-magnitude condensation in the Milky Way a few degrees east-northeast of Shaula. **Messier 6** is a magnitude dimmer than M7 but glows in a darker area 4° to the northwest. Known as the **Butterfly Cluster**, M6 contains many blue stars plus a prominent orange one in its eastern "wing." While M6 is 2,000 light-years from Earth, M7 is less than half that distance. If you're observing from the country, these clusters are dazzling in binoculars; you might even spot them with your bare eyes. Just bear in mind that southern Scorpius hugs the horizon at mid-northern locations.

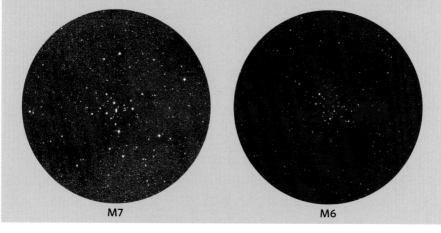

M7   M6

Sagittarius, *the Archer*

In legend, Sagittarius was a centaur, a remarkable creature that featured the head and arms of a man, plus the body and legs of a horse. Most centaurs were rude, drunken louts, and Sagittarius was no exception. An archer with an attitude, Sagittarius was determined to avenge the slaying of Orion by Scorpius (see page 24). In the sky we see Sagittarius stalking his quarry to the west, his bow and arrow aimed with deadly precision at the Scorpion's heart.

Can't picture the Centaur? Try the **Teapot,** a delightful asterism formed by eight 2nd- and 3rd-magnitude stars. Oddly, **Alpha (α)** and **Beta (β) Sagittarii** are not among them. Alpha and Beta are only 4th magnitude and are located so far below the Teapot, you won't see them if you're located north of about latitude 45° (the declination of Beta is –45°). However, if you can spot **β¹,** or **Arkab,** (perhaps with the aid of binoculars) you'll discover that it has a 4.3-magnitude companion, **β²,** directly under it.

We see the Teapot canted on an angle with a cloud of "steam" (the Milky Way) rising from its spout. The late *Sky and Telescope* columnist George Lovi once described a complete tea service all around Sagittarius: Below and to the right of the spout, the curving tail of Scorpius makes a large teacup. Above the Teapot's handle, a line of faint stars forms a teaspoon. Under the pot, a semicircle of faint stars in the constellation Corona Australis (identified on the

CELESTIAL TEAPOT: The "steam" forming above the "spout" of the popular Teapot is in reality the Sagittarius Star Cloud, a brilliant sector of the Milky Way that's well worth scanning with binoculars. Note the nebula M8 and the globular star cluster M22 nearby.

chart) outlines a tasty slice of lemon.

The center of our galaxy is located behind western Sagittarius. The region is a celestial wonderland of deep-sky treasures (opposite). Sagittarius is also a member of the zodiac, so expect to see planets tracking past the Teapot from time to time. The Sun reaches its southernmost point, called the *winter solstice*, near the Teapot's spout in late December.

▶ MISTY LAGOON

Look for **Messier 8**, the **Lagoon Nebula**, just above the spout of the Teapot. Like the Orion Nebula (see page 25), this lozenge-shaped object is an *emission nebula*, a vast cloud of hydrogen gas enshrouding a cluster of stars. Hold your binoculars steady to glimpse the family of young suns tangled in the eastern half of the Lagoon's misty web. M8 is 4,500 light-years from Earth (three times farther than M42) yet it is faintly visible to the unaided eye (away from city lights) as a 6th-magnitude brightening in the Milky Way.

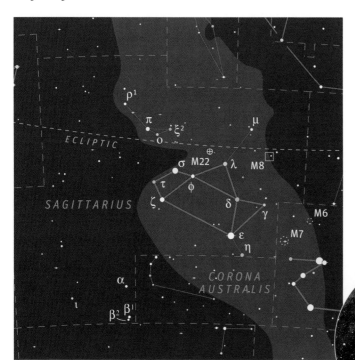

GORGEOUS GLOBULAR

A *globular cluster* is a dense ball of several hundred thousand extremely old stars. About 150 globulars are scattered throughout a halo centered on the hub of the Milky Way Galaxy. From our vantage point near the edge of the galaxy, the hub is in Sagittarius; consequently, a large number of globulars are visible there. The finest specimen in Sagittarius is **Messier 22**, the **Great Sagittarius Cluster**, which enjoys a picturesque setting near the "lid" of the Teapot. Only 10,000 light-years from Earth (close for a globular), M22 shines at 5th magnitude and is therefore visible to the eye in a dark sky. Ordinary binoculars show a fuzzy sphere, while tripod-mounted "giant" binoculars can partly resolve the cluster into stars.

Ophiuchus,
the Serpent Bearer

Ophiuchus is often identified with Aesculapius, a mythical healer who became the Greek god of medicine. Tradition holds that Aesculapius' knowledge was handed down to his descendants and, ultimately, to Hippocrates — a real person born about 460 BC and widely considered the father of medical science. Aesculapius possessed magical powers: he could cure the terminally ill and revive the dead. This miffed Pluto, god of the underworld, who complained that he was losing customers. In retaliation, Pluto convinced Zeus to equip the giant eagle, Aquila, with a thunderbolt and knock the doc out of commission. After the dirty deed was done, Zeus honored the doctor by placing him in the heavens as the constellation Ophiuchus. However, some say that Zeus just wanted Aesculapius nearby should the gods ever require his special healing powers.

The word Ophiuchus comes from a Greek phrase loosely meaning "to handle the serpent." Old star charts depict Aesculapius grasping a snake, whose venom he could transform into medicine

(see below). The star figure is huge (it ranks 11th among the constellations) but its shape is difficult to trace. Look for a stick-figure human of wide girth outlined by seven or eight medium-bright stars. The brightest is 2nd-magnitude **Alpha (α) Ophiuchi**, or **Rasalhague**, which marks the doctor's head.

Although the limbs of Ophiuchus aren't obvious, a "leg" of faint stars thrusts downward between Scorpius and Sagittarius. This leads us to a little known fact about Ophiuchus: the lower third of the constellation stretches southward right across the zodiac. Every December, the Sun spends nearly three weeks within the borders of Ophiuchus, while requiring just one week to cross Scorpius. The burly Serpent-bearer is the unofficial 13th sign of the zodiac!

Serpens, *the Serpent*

The celestial serpent writhes around the legs of its captor, Ophiuchus, who represents the fabled physician, Aesculapius. To Aesculapius, the snake was a magical creature that seemingly renewed itself by shedding its skin. The god of medicine took control of the serpent and conducted experiments with its poison. He discovered that snake venom could kill or cure, depending on the nature of its application. In time, the snake (and ultimately the *caduceus*, a staff with two intertwined serpents) became the universal symbol of medicine.

Modern-day celestial cartographers have divided Serpens into eastern and western sections. The two parts count as a single constellation, though it has two names. The constellation's western component, **Serpens Caput,** zigzags northward along Ophiuchus' western flank before terminating in a triangle of stars that marks the serpent's head. The snake's midsection — fleshed out on the charts of yesteryear — coils around the serpent bearer's legs but has no stars assigned to it. We pick up the tail of the constellation as it stretches northeastward from Ophiuchus's east side. Called **Serpens Cauda,** this long, straight segment fits snugly in a dark lane of the Milky Way called the Great Rift.

The brightest star in either part of the constellation shines modestly in Serpens Caput. It is orangey 2.6-magnitude **Alpha (α) Serpentis,** whose name, **Unukalhai,** means "the neck of the snake." See if you can spot the serpent's head, north of Unukalhai, which is formed by a triangle of 4th-magnitude stars. The triangle is topped by reddish yellow **Kappa (κ).** If you have binoculars, look for reddish 5th-magnitude **Tau[1] (τ[1]) Serpentis** just west of the head. Between Tau[1] and Kappa lies a scatter of at least eight 6th-magnitude stars, all easily picked up in binoculars.

Hercules, *Son of Zeus*

A tale of woe surrounds the Greek warrior Hercules. It begins with Zeus, who deceived his wife, Hera, by having an affair with a mortal woman. The product of this union was Hercules, a human boy of enormous strength and courage. The vengeful Hera tried to kill Hercules — without success — but ultimately caused him to suffer a spell of insanity during which he murdered his wife and family. Hercules' only hope for redemption lay in completing a series of onerous tasks known as the Twelve Labors. He survived the Labors but was later maimed by his overly jealous second wife who had smothered his clothes with a flesh-eating potion. Hercules finally had had enough. The hard-luck hero pleaded to his father for asylum in the heavens.

Hercules kneels with one foot — **Iota (ι)** — planted on Draco, a reptilian creature he slew during the second-last of his Twelve Labours. (Different legends about Draco appear opposite.) Hercules is oriented upside down. His head, which lies at the bottom of the constellation, is marked by **Alpha (α) Herculis,** or **Rasalgethi.** This red giant star fluctuates between 3rd and 4th magnitude — symbolic, perhaps, of the turbulent fate Hercules endured all his life. The Strongman's torso is a box of 3rd- and 4th-magnitude stars called the **Keystone,** which is easily found one third of the way from Vega to Arcturus.

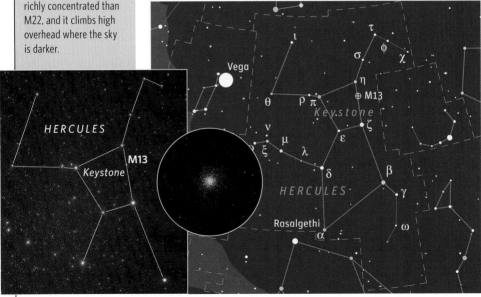

Patterns in the Sky

Draco, *the Dragon*

In Greek myth, Draco soldiered for the Titans who lost control of the universe in a war with the Olympians. The Dragon fared badly, getting thrown into the sky with such force he crashed into the north celestial pole. There, his writhing coils got caught in the axis of the world and the polar cold froze him solid.

The dragon theme was probably reworked from an ancient Mesopotamian creation myth. According to the *Enuma Elish,* the grotesque entity called Tiamat had ruled the primeval darkness before there was worldly form or substance. When Tiamat morphed into a dragon, the Babylonian hero Marduk battled it for control of the cosmos. Emboldened with special powers, Marduk slew the dragon and ripped it in two, using the halves of its body to fabricate heaven and Earth. He then created the stars and set them in motion. (A variation of this story appears on page 85.)

Draco's most famous starry member is 3.7-magnitude **Alpha (α) Draconis,** or **Thuban,** in the Dragon's back. Nearly 5,000 years ago, when the Earth's axis was oriented in a slightly different direction, Thuban was the pole star instead of Polaris. The Great Pyramid at Giza was designed to align with Thuban.

You'll find the Dragon curled part way around the Little Dipper. Draco's head, marked by 2nd- through 5th-magnitude stars, is a boxy asterism half the size of the nearby Keystone. Aim your binoculars at **Nu (ν) Draconis,** the faintest star in the head of Draco. Nu is an easy double that displays identical 5th-magnitude suns.

Summer Sky

And there revolves herself, image of woe,
Andromeda beneath her mother shining.
— Aratus

▶ Introducing the
Autumn Sky

Lengthening autumn nights are ideal for getting in a few extra hours of stargazing before bedtime. If you consult the seasonal star map inside the back cover of this book you'll find what's up in terms of autumn sky patterns.

The dozen autumn groups described in this chapter occupy center stage. Three zodiacal constellations — Capricornus, Aquarius, and Pisces — are rather dim, while a fourth (Aries) is small but easy. Cepheus, Cassiopeia, Andromeda, Perseus, Pegasus, and Cetus aren't uniformly prominent but together they tell a larger-than-life story. The legend of Andromeda and Perseus plays out every clear autumn night.

MISSION ANDROMEDA

Long ago, Andromeda was the lovely daughter of King Cepheus and Queen Cassiopeia of Ethiopia. Cassiopeia carelessly boasted that her beauty exceeded that of the Nereid sea nymphs in the court of the sea god, Poseidon. The Nereids implored Poseidon to neutralize the vain Queen. Poseidon responded angrily by flooding the coastal kingdom and dispatching a huge sea monster named Cetus to ravage all who lived there. Consulting an oracle, Cepheus learned that the only way he could save his people was to sacrifice his daughter to the voracious monster. Accordingly, Andromeda was chained to a rock by the ocean so that Cetus could devour her.

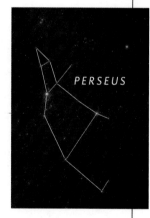

Enter Perseus. The plucky protagonist was returning from a dangerous adventure during which he had destroyed a horribly disfigured witch called Medusa. Merely setting eyes on Medusa would turn anyone instantly to stone but the young warrior had cleverly decapitated the woman by looking at her reflection in his shield and had stuffed her head in a sack. Now Perseus had chanced upon Andromeda in her moment of peril. Our hero knew what to do: he pulled Medusa's severed head out of his sack and thrust it at the advancing sea monster. Cetus let out an awful cry then morphed into a huge rock. Perseus freed the terrified princess and carried her safely homeward.

Finding Your Way
in the Autumn Sky

Your celestial signpost for autumn is an asterism called the **Great Square of Pegasus.** Picturing a horse among the stars of Pegasus is somewhat challenging, but the Great Square inside the constellation is easy. Even so, its name is slightly misleading. Firstly, it isn't a perfect square. The big box measures about 14° high and 16° wide. Secondly, its stars are bright but not brilliant. None of them quite match the magnitude of Polaris, the North Star. Nevertheless, identifying the Great Square is a straightforward exercise.

The Square's southwestern corner is marked by 2.5-magnitude **Alpha (α) Pegasi,** or **Markab.** At the northwestern corner is 2.4-magnitude **Beta (β) Pegasi,** or **Scheat.** Scheat is a red giant whose color is obvious in binoculars and whose brightness is slightly variable. At the southeastern corner is 2.8-magnitude **Gamma (γ) Pegasi,** or **Algenib.** The northeastern corner is marked by 2.1-magnitude **Alpha Andromedae,** or **Alpheratz.** Although Alpheratz technically belongs to Andromeda, not Pegasus, the star does double-duty as the brightest member of the Great Square.

The Great Square stands out mainly because no other bright star pattern is nearby. It works as a signpost, in part, because the Square's horizontal sides are oriented east-west and the vertical sides are aligned north-south. This square-rigged asterism can help you navigate the entire autumn sky.

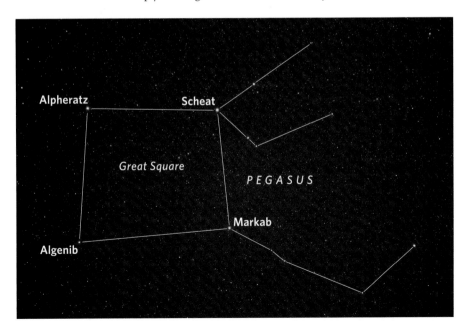

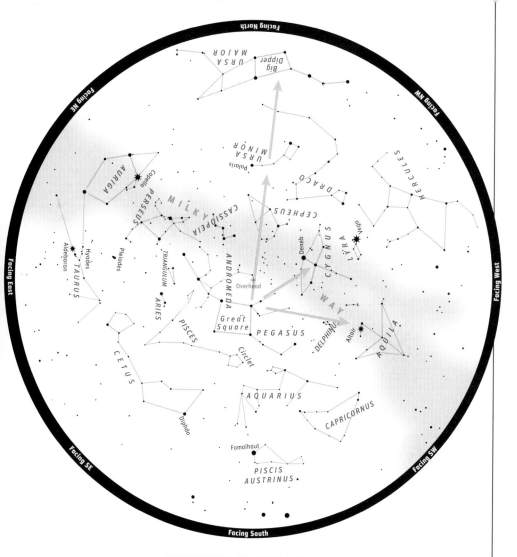

WESTWARD, BACK TO SUMMER

Let's begin our orientation by following the Great Square of Pegasus to a few summer stragglers. A line from Alpheratz across the top of the Great Square through Scheat aims westward toward **Aquila,** fairly low in the west-southwest. A line from Algenib drawn diagonally across the Great Square through Scheat points west-northwestward toward **Cygnus** and **Lyra,** which hang high in the western sky. Below them is Hercules, which sprawls just above the northwestern horizon.

Before we explore further, let's see if we can spot the **Big Dipper.** You might be having trouble locating the Dipper because it's low in the north where it can be obscured by houses and trees. Provided the Dipper is not actually below your horizon, the Great Square will guide you to it. Simply draw a line from Markab upwards along the western side of the Great Square, through Scheat, and continue northward past the zenith into the northern sky. The line goes through northern **Cepheus,** passes by **Polaris,** and then descends into the bowl of the Big Dipper near the northern horizon.

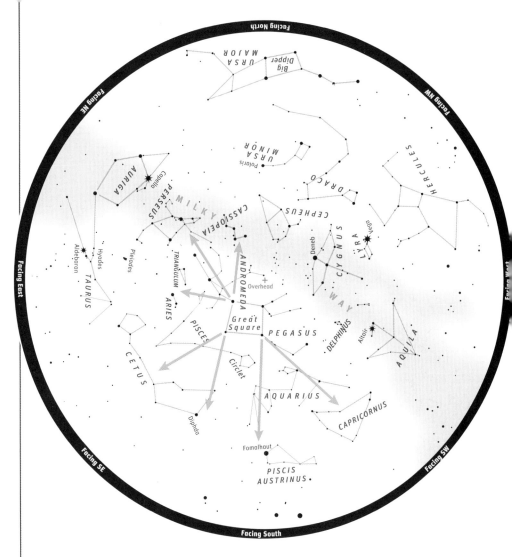

GETTING AUTUMN SQUARED AWAY

The Great Square points to all the autumn constellations described on pages 76 to 87. A line from Alpheratz diagonally across the Great Square through Markab aims south-

westward past **Aquarius** into **Capricornus.** The same line extended in the opposite direction cuts across **Andromeda** before continuing northeastward into **Perseus.** A line from Scheat diagonally across the Square through Algenib aims southeastward past **Pisces** into the heart of **Cetus.**

A line southward from Algenib runs past Pisces toward **Diphda** in Cetus. The same line drawn in the opposite direction touches **Cassiopeia.** A line extended southward from Markab travels through Aquarius to the bright star **Fomalhaut** in **Piscis Austrinus.**

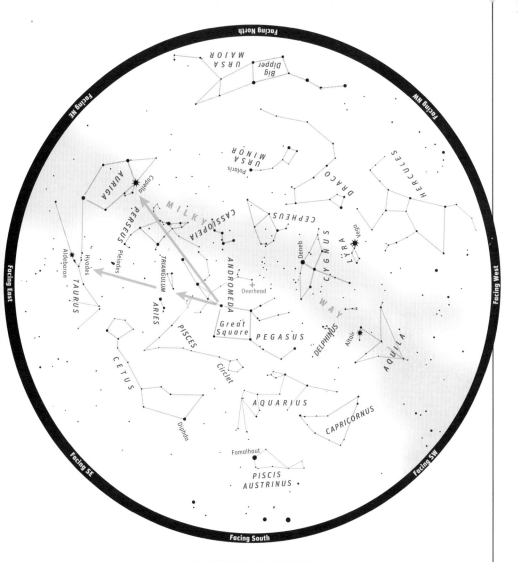

EASTWARD, TO WINTER

Allow the Great Square to direct you to a couple of the rising constellations of winter. A line drawn eastward from Alpheratz cuts between **Triangulum** and **Aries.** Extending that line farther eastward brings us to bright red Aldebaran in **Taurus** just above the eastern horizon. A line drawn northeastward from Markab cuts diagonally across the Square past Alpheratz then passes through Andromeda and Perseus all the way to brilliant Capella in **Auriga** above the northeastern horizon. Taurus and Auriga are the advance guard for the brilliant winter sky patterns described on pages 24 to 33.

AUTUMN EXTRAS

Three minor constellations in the autumn sky have not been included in this chapter. They are **Equuleus**, the Little Horse (a tiny, dim group between Pegasus and Delphinus); **Lacerta**, the Lizard (an ill-defined pattern in the gap bordered by Pegasus, Cygnus, and Cepheus); and **Sculptor**, the Sculptor (a mid-size but very faint constellation east of Piscis Austrinus).

Cepheus, *Father of Andromeda*

Cepheus, the legendary King of Ethiopia, was the father of Andromeda and husband of Cassiopeia, the vain Queen who incurred the wrath of the sea god, Poseidon. (Their story appeared on page 71.) Some old charts show Cepheus with his hands upraised, as if appealing to the gods for mercy. On other charts King Cepheus is guarding the north celestial pole, his left foot planted with authority on Polaris. However, the royal pedigree of this northern constellation predates classical mythology. To the astronomer-priests of ancient Egypt, Cepheus represented Khufu, the builder of the Great Pyramid. In Babylon, Cepheus was considered a descendant of the sky-god Enlil who ruled the circumpolar heavens.

The five main stars of Cepheus range between magnitude 2.5 and 3.5. Connecting the dots produces a stick-figure house with a pointy roof. In autumn the house hangs inverted, its roof aiming downward past the North Star. Glowing dimly near the southeastern corner of the constellation is the famous variable star **Delta (δ) Cephei**. An aging star that pulsates with rhythmic regularity, Delta's light output cycles between magnitude 3.5 and 4.4 every 5.4 days. It forms a tiny isosceles triangle with 3.4-magnitude **Zeta (ζ) Cephei** and 4.2-magnitude **Epsilon (ε) Cephei**. When Delta is at

A STAR THAT PULSES: The variations of Delta Cephei are easy to monitor by comparing the star with neighboring Zeta and Epsilon. Nearby, Mu Cephei glows with a lovely deep red color.

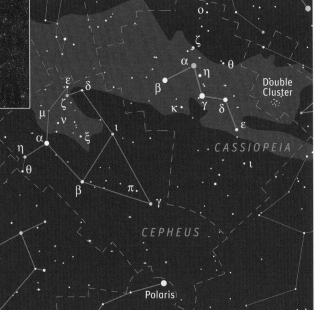

maximum it's almost as bright as Zeta; at minimum it's slightly fainter than Epsilon. Binoculars show all three stars; they might also reveal a bluish, 6th-magnitude companion star next to deep yellow Delta.

While in the area, aim your binoculars 6° west of Delta Cephei at a celestial stoplight called **Mu (μ) Cephei.** Nicknamed the Garnet Star, this red supergiant fluctuates erratically between 3rd and 5th magnitude over a period of about two years. You might not notice the variability of the Garnet Star, but you can't miss its ruby red glow!

Cassiopeia,
Mother of Andromeda

Cassiopeia, the beautiful but conceited mother of Andromeda (and wife of King Cepheus), displays little of her feminine form in the sky. Like Cepheus, Cassiopeia is a circumpolar constellation with a pattern of five main stars. However, the stars of Cassiopeia are somewhat brighter (ranging between magnitude 2.2 and 3.4) and they form an eye-catching letter of the alphabet — *two* letters if you observe the constellation in spring as well as in autumn. In spring, when Cassiopeia swings near the northern horizon, the pattern resembles a giant letter W. In autumn, the constellation arcs high overhead and flips over to become the letter M.

Be it a W or an M, Cassiopeia's alphabetical outline is more suggestive of an angular royal throne than the Queen herself. That throne had better come with seat belts because for half the year it hangs upside down and the poor woman must hold on for dear life. Her predicament is explained in a postscript to the legend of Perseus and Andromeda. Clearly, the gods were not amused with Cassiopeia's unrestrained vanity. As a lesson in humility, they placed the loquacious lady in the far northern sky where she is forced into her undignified orientation for several months at a time. The Queen's loss is our gain: Because Cassiopeia is located within 30° of the north celestial pole, the constellation is visible every clear night of the year.

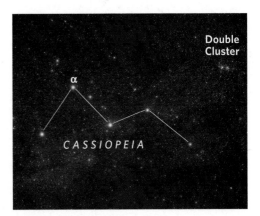

STARRY M: The Milky Way sprawls across Cassiopeia. Note that four of the constellation's five main stars shine with a blue-white color. The exception is 2.2-magnitude **Alpha (α) Cassiopeiae,** which glows with an orange hue. The double-dollop of stars in the upper-right corner is the famous **Double Cluster** in adjacent Perseus (see page 81).

Pegasus, *the Winged Horse*

Pegasus, the magnificent flying horse, was born under trying circumstances. According to one legend, he materialized from a drop of blood shed by the winged witch Medusa whom Perseus had just slain (for the gory details, turn the page). It seems ironic that pearl-white Pegasus should inherit his angelic wings from such a horrific creature.

An earlier legend links the horse to a different rider. The warrior Bellerophon engaged Pegasus to help him slay a fire-breathing dragon called Chimera. Emboldened by his victory, Bellerophon implored his winged steed to soar upward toward Mt. Olympus — and heaven itself. This ill-mannered maneuver infuriated Zeus who caused the horse to bolt, tossing Bellerophon to an undignified end. Consequently, it was Pegasus — not Bellerophon — that was awarded immortality among the stars.

Identifying the celestial stallion is a bit tricky. Although Pegasus is the seventh largest constellation in the sky, the winged horse it symbolizes flies upside down. Moreover, the star pattern depicts only the front half of the horse. Its main body is outlined by the four medium-bright stars of the **Great Square,** which was described on page 72. The horse's front legs extend from the Square's northwestern corner — marked by reddish **Beta (β) Pegasi** or **Scheat** — as two rows of 3rd- and 4th-magnitude stars. The neck protrudes from **Markab — Alpha (α) Pegasi** — on the southwestern corner as another row of 3rd- and 4th-magnitude

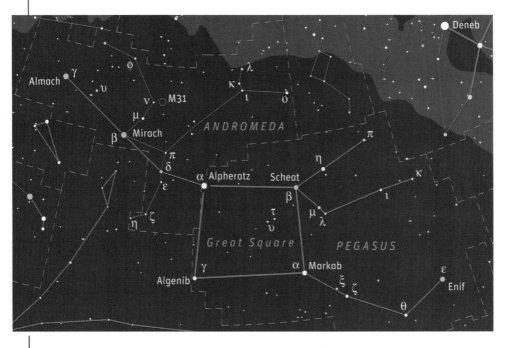

stars. The head of the horse is represented by golden-hued, 2.4-magnitude **Epsilon (ε) Pegasi,** or **Enif** ("nose").

Although the Great Square occupies about 200 square degrees of sky, it contains few naked eye stars. None of them are brighter than magnitude 4.4. How many stars can you detect inside the Great Square?

Andromeda,
Daughter of Cassiopeia

When we last saw Princess Andromeda on page 71, she had been rescued from a marauding sea monster by the brave hero Perseus. Gazing skyward at her constellation, it can be difficult to recognize the famous femme fatale because the star pattern makes no obvious female figure. But Andromeda is famous for something else, as we shall see.

At the apex of the constellation is 2.1-magnitude **Alpha (α) Andromedae,** or **Alpheratz,** which establishes the girl's head (Alpheratz also marks the northeastern corner of the Great Square). From Alpheratz, a line of three stars runs northeastward. They are 3.3-magnitude **Delta (δ)**, 2.0-magnitude **Beta (β)**, or **Mirach,** and 2.1-magnitude **Gamma (γ)**, or **Almach** (a spectacular binary in telescopes). North of, and roughly parallel to, that line is a trio of dimmer stars whose middle marker is 3.9-magnitude **Mu (μ).** Mu is the gateway to our prize. A line from Beta to Mu, extended its own length northwestward past 4.5-magnitude **Nu (ν)**, points to the **Andromeda Galaxy.**

M110

M32

▸ THE ANDROMEDA GALAXY

Located 2.5 million light-years from Earth, the **Great Andromeda Galaxy,** or **Messier 31,** is the most distant object most people can see with their naked eyes. Yet M31 is the closest major spiral galaxy to us. Its 4th-magnitude glow is easy to identify so long as your sky is reasonably dark. M31 is oriented nearly edge-on to our line-of-sight, so it appears as a cigar-shaped cloud. What you see in binoculars depends on your sky conditions. From the city, only the bright nuclear region shines through. Under a country sky, the central condensation grows slender "wings" that span at least 2° of sky. In the photo, note Andromeda's 8th-magnitude satellite galaxies, **M32** and **M110,** which show in telescopes.

ANDROMEDA

M31

β

α

Perseus, *Rescuer of Andromeda*

Perseus was born following an affair between Zeus and a mortal woman named Danae. Danae raised her child in the court of King Polydectes of Seriphos. Polydectes eventually proposed to Danae but she rejected him. Deeply offended, Polydectes turned against Perseus. He ordered the young man to present the royal court with the head of the Gorgon Medusa. A hideous witchlike woman with wings, clawed feet, and snake-infested hair, Medusa's gaze would turn any mortal instantly into stone. Fortunately, when our hero found Medusa, he eyed her safely via the reflection in his shield, then decapitated her. Perseus grabbed the severed

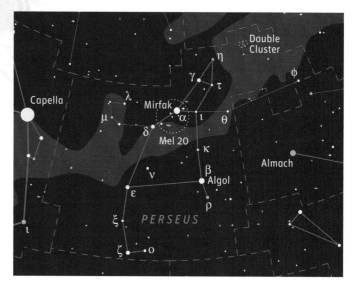

MELOTTE 20

The band of the Milky Way runs through the center of Perseus. A conspicuous patch of Milky Way surrounding 1.8-magnitude **Mirfak** is actually an elongated, loosely bound star cluster called the **Alpha Persei Association,** or **Melotte 20.** After Mirfak, the dozen brightest members of Melotte 20 shine between 4th and 6th magnitude and are arranged in a serpentine pattern

almost 5° long. Many fainter stars are scattered between Mirfak and 3.0-magnitude **Delta (δ) Persei** to Mirfak's southeast (Delta itself is not a member). Viewed from a country observing site, Melotte 20 is a naked-eye object and is impressive in binoculars. The slightest optical aid will reveal it in a city sky. The Alpha Persei Association is about 600 light-years away.

THE DOUBLE CLUSTER

The famous **Double Cluster** beckons from its location in northernmost Perseus. If sky conditions are good, this glittering showpiece is visible to the eye as an elongated "fuzz" in the Milky Way about one third of the way from 3.8-magnitude **Eta (η) Persei** to 2.7-magnitude **Delta (δ) Cassiopeiae** (see the photo on page 77). Even from the city, binoculars show the Double Cluster as side-by-side clumps of haze speckled with a few pinpoints of light. Together the objects cover almost 1° of sky in an east-west orientation. The eastern cluster, **NGC 884,** is magnitude 6.1 and includes some reddish suns among

dozens of blue-white members. The western cluster, **NGC 869,** shines at magnitude 5.3, is more concentrated, and contains

a slightly larger number of stars. The clusters are 7,600 and 6,800 light-years away respectively.

head and, following a brief time-out to rescue Princess Andromeda, returned to Seriphos. There, Perseus confronted Polydectes and dangled the gruesome Gorgon in front of him. The King was literally petrified.

The dozen or so stars in Perseus' gently curving pattern shine between 3rd and 4th magnitude, with two exceptions. The first is yellowish, 1.8-magnitude **Alpha (α) Persei,** or **Mirfak** ("the elbow"). The other is blue-white **Beta (β) Persei,** or **Algol** ("the demon"), which symbolizes Medusa's head. Like Beta Lyrae (page 59), Algol is an *eclipsing binary.* The Demon Star usually shines at magnitude 2.1 but every three days it dims to magnitude 3.4. The light is low for about 10 hours, then it returns to normal. (See page 95 for a Web site that predicts Algol's minimum brightness.) Keep watch on Algol by comparing it to 2.9-magnitude **Epsilon (ε) Persei** and 2.1-magnitude **Almach** in neighboring Andromeda. When the Demon Star is at maximum light, it is as bright as Almach; at minimum, it's dimmer than Epsilon Persei.

Finally, check out the semi-regular variable star **Rho (ρ) Persei** 2° south of Algol. Rho wavers between magnitude 3.3 and 4.0. The view in binoculars is pleasing, for Algol is pure white, while Rho is reddish-orange.

Aries, *the Ram*

Aries was a golden flying sheep sent by the gods to save a young prince and princess from their wicked stepmother. As the ram tried to airlift the children to safety, the girl fell to her death but the boy held on. The ram landed his passenger in Colchis by the Black Sea. In gratitude for his rescue, the prince sacrificed the ram to Zeus, then gave the gilded hide to the King of Colchis who nailed it to a tree. The Golden Fleece was eventually retrieved by Jason and the Argonauts during their illustrious voyage across the eastern Mediterranean.

The Ram also figures in the "Titanomachia," the war in which the Olympian gods fought the Titans for control of the universe. At one point Zeus and his cohorts were forced to retreat to Egypt. While resting beside the Nile River, they were ambushed by the enormous Titan serpent, Typhon. Zeus disguised himself as a ram and, with the aid of a "sea goat" (page 87), barely escaped. The ram later became a constellation of the zodiac.

Greek skywatchers noted that the Sun passed through Aries each year during the *spring*, or *vernal equinox*. The golden ram thus

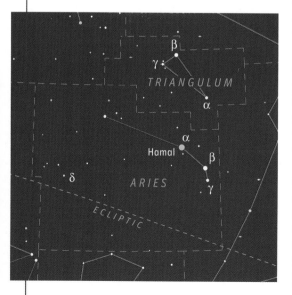

became a symbol of the Sun's revitalization after months of winter. Observers began reckoning celestial longitude (called right ascension) from the vernal equinox. This key spot, where the sky's prime meridian crosses the vernal equinox, was dubbed the First Point of Aries. The term is still in use today even though the vernal equinox is now in western Pisces.

Aries' angular outline is easy to identify east of Pegasus. The Ram's brightest star is 2.0-magnitude **Alpha (α) Arietis,** or **Hamal,** which shines with an orangey tint. **Beta (β)** is magnitude 2.6, while **Gamma (γ)** is only magnitude 3.9. Look for bright planets hugging the ecliptic just below these stars.

Triangulum, *the Triangle*

Gazing up at the constellation Triangulum, the ancient Greeks saw their capital letter Delta (Δ) written in the stars. The calligraphy is somewhat flawed because Triangulum's star pattern, which resembles a long isosceles triangle, is too slender to make a convincing Δ. As if to compensate for this, the Greek chart makers pictured the constellation as an equilateral triangle! Similarly, Egyptian sky watchers considered Triangulum a symbol of the Nile River Delta. Nobody seemed to mind that the broad delta on the ground didn't match the much narrower triangle in the sky.

The astronomers of the Roman Empire associated Triangulum with the island of Sicily. According to legend, Ceres, the Roman goddess of agriculture, admired Sicily's fertile landscape so much she lobbied the head god, Jupiter, to have a likeness of the island placed in the heavens. Once again, the star figure was somewhat narrower than the geography it represented. The perceived connection between Triangulum and Sicily surfaced once more on New Year's Day, 1800. On that date, the director of Sicily's Palermo Observatory, Giuseppi Piazzi, discovered the first asteroid in the solar system. Because it was found near Triangulum, he asked that it be named Ceres in honor of Sicily's patron goddess.

The tenth smallest constellation in the sky, Triangulum is hardly a prize-winning pattern. Its brightest star is 3.0-magnitude **Beta (β) Trianguli,** while the vertex is marked by 3.4-magnitude **Alpha (α).** However, the triangle's eastern corner is worth a closer look. Excellent eyesight (or binoculars) will reveal that it is established by not one, but three stars. The most obvious is 4.0-magnitude **Gamma (γ).** A 4.8-magnitude star shines above Gamma and a 5.3-magnitude star is below. Finally, aim your binoculars at Gamma itself. If its glare isn't too strong, you'll spot an 8th-magnitude companion shining immediately below Gamma.

Pisces, *the Fishes*

Ancient stargazers imagined V-shaped Pisces as a pair of fishes joined by a long ribbon. This curious arrangement surfaced in Roman sky lore, which identified Pisces as the goddess Venus and her son Cupid (in Greek mythology, Aphrodite and Eros). In one legend, they were attacked by a huge monster. Desperate to escape, Venus and Cupid jumped into the sea and transformed themselves into fishes. So that they wouldn't become separated, mother and son connected themselves with the ribbon.

A knot in the ribbon is marked by 3.8-magnitude **Alpha (α) Piscium,** or **Alrescha,** which means "cord." With the exception of 3.6-magnitude **Eta (η)** and 3.7-magnitude **Gamma (γ),** the other stars in Pisces are fainter than Alpha. Consequently, even when the sky is perfectly clear, city observers will probably have trouble identifying this sprawling constellation. Its most distinctive — albeit dim — feature is an asterism called the **Circlet,** an elongated ring of 4th- and 5th-magnitude stars that outlines the western fish just below the Great Square of Pegasus. The Circlet is 7° wide; see if it fits in your binoculars' field of view. Any binoculars will show two modest trea-

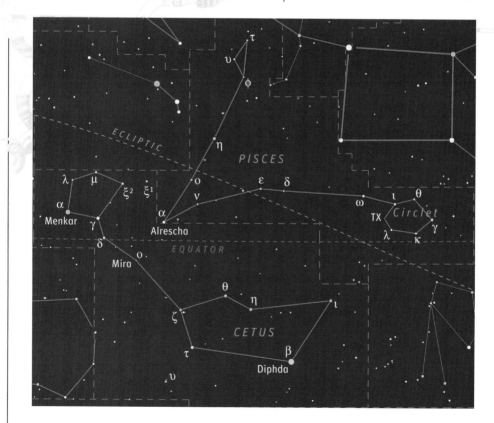

Patterns in the Sky

sures in the Circlet. One is 4.9-magnitude **Kappa (κ) Piscium,** which has a wide 6.3-magnitude companion. The other is **TX Piscium,** a deep red variable star that hovers around 5th magnitude.

Pisces is accompanied by several other "watery" constellations, such as Cetus, Aquarius, and Capricornus. Ancient Middle Eastern observers would experience their rainy seasons when the Sun passed through this sloshy part of the sky every winter and spring. The vernal equinox, which marks the first day of spring, has been drifting across Pisces for more than 2,000 years. Currently the equinox is located in western Pisces, under the Circlet.

Cetus, *the Sea Monster*

Cetus is strange even by sea monster standards. The mythmakers of classical Greece conjured up a whale-like amphibian that, depending on the storyteller, featured a hound's head, stubby forelegs, webbed feet with claws, and a lashing, trident tail. Wise astrologers warned that whatever its shape, Cetus represented the very essence of anarchy and chaos.

Cetus, and the fearsome theme it represents, greatly predates Greek mythology. Mesopotamian skywatchers saw Cetus, plus the northern constellation Draco (page 69), as symbols of the grotesque Tiamat, a sea monster or dragon that ruled the primordial waters before the universe took form. According to the Babylonian creation epic *Enuma Elish,* Tiamat was destroyed by a hero named Marduk, who fashioned the cosmos from the creature's hide. When he designed the constellations, Marduk included Cetus and Draco as a reminder of humanity's valiant struggle against evil.

Although Cetus is the sky's fourth largest constellation, it contains few prominent stars. But that's not to say they aren't interesting. For example, you may (or may not!) spot **Omicron (o) Ceti,** or **Mira** ("the Wonderful"). Located 220 light-years away, Mira is the most famous long-period variable star in the sky. The brightness of this unstable red giant changes by a factor of more than 200, usually from 3rd to 9th magnitude, over an average period of 332 days. If you're observing from an urban location, most of the time Mira won't be visible without binoculars or a small telescope.

Northeast of Mira, the monster's head is outlined by a large pentagon of five mostly faint stars. Its brightest member, reddish 2.5-magnitude **Alpha (α) Ceti,** or **Menkar** ("nose"), is accompanied by a bluish, 5.6-magnitude companion (visible in binoculars). The hindquarters are formed by a half-dozen somewhat brighter stars including 2nd-magnitude **Beta (β) Ceti,** also known as **Diphda** or **Deneb Kaitos** ("tail of the whale").

Aquarius, *the Water Carrier*

On page 60, I left a young Greek shepherd named Ganymede literally hanging in the air. The boy was carried by the eagle Aquila to Mount Olympus where he spent the rest of his days as the official wine pourer for the gods. In recognition of his valuable service, Zeus placed Ganymede in the heavens as the constellation Aquarius.

The Water Carrier we know — Aquarius himself — dates back to Babylonian times, when the constellation represented a man pouring water from a jar. Although Aquarius ruled "the water," he was reviled when he hosted the Sun during the wettest part of winter. This period was considered the "curse of rain" by the Chaldeans, who complained that the water-carrier's overflowing jars were responsible for the epic flood described in the Gilgamesh.

By contrast, Egyptian skywatchers venerated Aquarius as a symbol of good fortune. They imagined him emptying his buckets into the Nile River, thereby triggering the Nile's life-sustaining flood each spring. Thankful farmers responded by giving the names **Sadalmelik,** "the Lucky One of the King," and **Sadalsuud,** "the Luckiest of the Lucky," to **Alpha (α)** and **Beta (β) Aquarii,** both magnitude 2.9. Most of the other stars in Aquarius are 4th magnitude and fainter. Chief among them, just east of Alpha, are four dim stars forming a Y-shaped asterism called, fittingly, the **Water Jar.**

Aquarius is the second largest constellation in the zodiac. The planet Neptune was discovered in Aquarius in 1846 at the Berlin Observatory following calculations by the English astronomer John Couch Adams and the French mathematician Urbain Leverrier.

Piscis Austrinus, *the Southern Fish*

Piscis Austrinus was the legendary parent of the two fishes comprising the constellation Pisces. A modest star pattern, Piscis Austrinus swims in relative obscurity beneath Aquarius. The Southern Fish is usually depicted with its mouth open, gulping the long stream of liquid flowing from Aquarius' overflowing Water Jar. This unending ingestion of heavenly water was interpreted as a heroic duty that helped Piscis Austrinus save the world from the Great Flood.

The gaping front end of Piscis is marked by 1.2-magnitude **Alpha (α) Piscis Austrini,** or **Fomalhaut,** the "Fish's Mouth." Fomalhaut is the only first-magnitude star among the autumn sky patterns, and shines nearly 30° below the celestial equator; northern observers may have to peer between trees and houses to spot it.

Capricornus, *the Sea Goat*

The origin of this strange hybrid of fish and goat can be traced to the "Titanomachia" legend I related on page 82. The Olympian gods were reclining along the banks of the River Nile when the monster Typhon attacked them. The god Pan, a musician who was part man, part goat, dove into a nearby river and tried to become a fish so he could swim away. While the transformation didn't quite work — Pan's hindquarters developed a powerful tailfin while the rest of him remained a goat — he made good his escape. Pan then turned back to help rescue Zeus. In gratitude, Zeus placed Pan in the zodiac as the constellation Capricornus.

Although Capricornus is the smallest constellation in the zodiac, it was venerated by the ancients. More than 4,000 years ago, the Sun occupied Capricornus as it passed through the winter solstice each year. Perhaps in recognition of the Sun's rebirth following the solstice, Capricornus was sometimes viewed as the gateway through which human souls passed when they ascended to heaven.

The constellation's wedge-like pattern is dimly outlined by a dozen 3rd- and 4th-magnitude stars. However, **Alpha (α)** and **Beta (β) Capricorni,** less than 2½° apart, deserve special mention. Alpha Capricorni, or **Algedi,** is a wide double consisting of 3.6- and 4.3-magnitude stars divisible by the naked eye (and certainly in binoculars). This pair only looks related; the brighter star is 109 light-years away while the dimmer one is about six times more distant. Beta Capricorni, or **Dabih,** is an easy binocular double with 3rd- and 6th-magnitude components.

▶ Go South!

Centaurus and Crux are two of the finest constellations of the deep southern hemisphere. Unfortunately, if you live in the northern United States, you can spot only the head of Centaurus poking above the horizon every spring (see the spring star map inside the front cover), and you can't see Crux at all. But if you journey to southern Florida, Mexico, or Hawaii, you can watch both constellations clear the treetops.

Centaurus, *the Centaur*

To the skywatchers of ancient Greece, Centaurus represented a centaur (half man, half horse) named Chiron. Unlike the warring centaur Sagittarius (see page 64), Chiron was a teacher who schooled generations of mythic Greek heroes in the arts and sciences. Chiron met an untimely end when his former student, Hercules, accidentally struck him with a poison arrow. An immortal, Chiron knew the toxin would doom him to eternal suffering. He appealed to Zeus to release him and subsequently he was elevated to the heavens. Old charts depict the thankful centaur sacrificing an animal (the constellation Lupus) to his Olympian peers.

Centaurus' huge pattern contains nearly two-dozen stars, most of them between 2nd and 4th magnitude. Its southeastern corner features two brilliant stars. One is –0.3-magnitude **Alpha (α) Centauri,** or **Rigil Kentaurus,** a golden yellow binary only 4.4 light-years away. Alpha also harbors a wide, 11th-magnitude star called **Proxima.** Located ¼ light-year closer to us than Alpha, Proxima has the honor of being the nearest star to the Sun. West of Alpha is bluish, 0.6-magnitude **Beta (β) Centauri,** or **Hadar,** which is more than 500 light-years away. These two luminaries point westward to the famous Southern Cross (see next page).

AMAZING OMEGA:
The best globular cluster in the heavens is Omega (ω) Centauri. Although 17,000 light-years away, Omega appears as a 4th-magnitude fuzzy patch 36° directly south of 1st-magnitude Spica in Virgo. The elliptical cloud, twice as wide as the full Moon, is unmistakable in binoculars and resolves easily in a telescope.

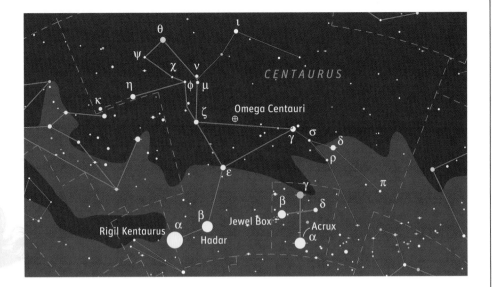

Crux, *the Cross*

Crux, the Cross (popularly known as the Southern Cross), is the smallest constellation in the heavens. Northerners chuckle that the 4° × 6° figure would fit inside the bowl of the Big Dipper. Nevertheless, the Cross's distinctive pattern is so entrenched in the public mind that it decorates the flags of several Southern Hemisphere nations. Crux lies beneath the Centaur, from which it was calved — and renamed — by chart makers in the 16th century. Yet the Cross was recognized at least 2,000 years ago. Christian astronomers saw the asterism above the southern horizon of Jerusalem.

The Cross is anchored at the bottom by blue-white, 0.7-magnitude **Alpha (α) Crucis**, or **Acrux**, 321 light-years away. Binoculars show a 4.8-magnitude companion star and Acrux itself is a gorgeous binary in telescopes. At the top of the Cross is reddish, 1.6-magnitude **Gamma (γ)**, or **Gacrux** (88 light-years), plus its wide, 6.5-magnitude companion. On the left is bluish **Beta (β)**, or **Mimosa** (353 light-years), a subtle variable star whose light wavers between magnitude 1.2 and 1.3. On the right is blue-white, 2.8-magnitude **Delta (δ)**, 364 light-years distant. The Cross's "fifth star," 3.6-magnitude **Epsilon (ε)**, exudes an orangey tint and is 228 light-years away.

Crux is a stunner in binoculars. The stretch of Milky Way in and around the constellation is peppered with clusters and nebulas. In stark contrast, the **Coalsack** immediately southeast of the Cross is a region apparently devoid of stars. But this is an illusion. The Coalsack is a *dark nebula* — a thick cloud of interstellar dust that hides more distant suns. The Coalsack is an arresting sight because of the way it seemingly erases a portion of the Milky Way.

CELESTIAL JEWELS:
Colorful stars highlight the Southern Cross. Look next to Beta Crucis for a compact open cluster called the Jewel Box. Although nearly 5,000 light-years away, the **Jewel Box** appears as a 4th-magnitude fuzzy star. Steadily held binoculars will resolve the brightest of its multi-colored gems. Note also the murky **Coalsack.**

Farther South

If you take your subtropical vacation during the December holiday period, two brilliant stars of the southern hemisphere will catch your attention. One is 0.5-magnitude **Achernar,** at the south end of Eridanus, the River. Achernar is 40° southeast of 1st-magnitude Fomalhaut (page 86). The other luminary is –0.6-magnitude **Canopus,** in western Carina, the Keel. Canopus is 36° south, and slightly west, of blazing Sirius (page 32). Viewed from Florida, Mexico, or Hawaii around 10 p.m. (local time) in late December, Achernar is sinking in the south-southwest while Canopus is rising in the south-southeast. Should you travel into the southern hemisphere, you can use these stars as stepping stones to several additional sky targets.

The Large Magellanic Cloud

Our Milky Way Galaxy possesses two satellite galaxies called the Magellanic Clouds (after the 16th-century navigator Ferdinand Magellan). Hovering within 20° of the south celestial pole, the **Large Magellanic Cloud,** or **LMC,** sprawls across southern Dorado, the Goldfish, and northern Mensa, the Table. The **Small Magellanic Cloud,** or **SMC,** is located in Tucana, the Toucan. These host constellations have colorful names but display no obvious star patterns; their most outstanding features are the Magellanic Clouds.

The Large Magellanic Cloud is roughly 25,000 light-years across, or about one-quarter as wide as its parent galaxy. Although 160,000 light-years away, the LMC is visible in a country sky as an elongated nebulosity approximately the size of the Southern Cross. Look for it 19° south-southwest of the brilliant star Canopus and 60° south of Orion. (Both Orion and the LMC cross the meridian before midnight in late December.) Slowly scan the bar-like cloud with binoculars or a telescope and you'll be confronted with innumerable faint stars, small clusters, and faint nebulas.

MAGELLANIC MAGIC: In this shot of the Large Magellanic Cloud, the pink patch is the **Tarantula Nebula.** The spider-like Tarantula is similar to the Orion Nebula (page 25) except that it is more than 100 times farther away. Despite this, the Tarantula appears as a 4th-magnitude patch almost as wide as the full Moon.

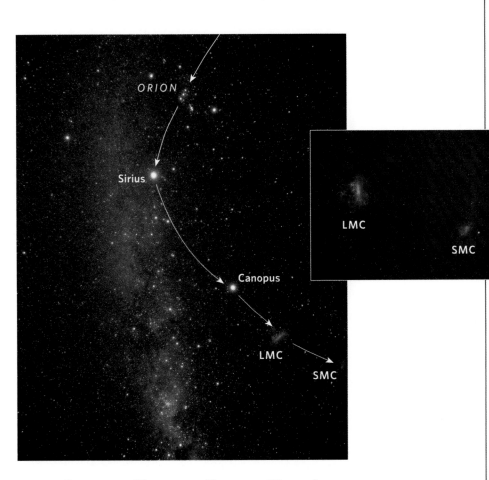

The Small Magellanic Cloud

The Small Magellanic Cloud, or **SMC,** is visible to the eye as an oblong haze about 3½° in its longest dimension. Almost 200,000 light-years away, the SMC is smaller and fainter than its sibling, the LMC, but is still easy to find 17° south, and slightly west, of the brilliant star Achernar (and far below the Great Square of Pegasus). In late December, the SMC crosses the meridian at nightfall; catch it early before it gets too low. The little cloud displays less structure than the big cloud, but nonetheless contains a variety of small clusters and nebulas. It's well worth exploring with binoculars or a small telescope.

MORE MAGIC: The Small Magellanic Cloud occupies the center of this image. The starry sphere near the top is the globular cluster **47 Tucanae.** Almost twice as wide as the full Moon, "47 Tuc" is a 4th-magnitude blob seemingly near the SMC. The cluster is actually 16,000 light-years from Earth. 47 Tucanae is admired for its incredibly dense concentration of stars. Binoculars hint at this glory, showing a diffuse glow around a starlike nucleus.

▶ Glossary

asterism: An easily recognized star pattern, usually with a popular name, formed from stars in one or more constellations. The most famous asterism in the sky is the Big Dipper in Ursa Major. See also **constellation.**

astronomy: The theoretical and observational study of the universe and its contents. The physical science of astronomy has nothing to do with the practice of *astrology*, which suggests a connection between the motions of solar system bodies and human traits or behavior.

binary star: Two stars held in orbit around each other by their mutual gravitational attraction. See also **eclipsing binary** and **double star.**

celestial poles: The *north celestial pole*, analogous to the North Pole of the Earth, is located at 90° north declination on the celestial sphere. The *south celestial pole* is at 90° south declination.

celestial equator: Divides the celestial sphere in half and is analogous to the Earth's equator.

celestial sphere: A mental construct in which the visible sky is pictured as a vast, hollow sphere turning around the stationary Earth. The sphere rotates westward, carrying every celestial object with it.

circumpolar star: A star that, viewed from a particular latitude, never sets below the horizon. The closer to the Earth's North Pole that an observer is located, the greater the fraction of his or her sky that is circumpolar.

constellation: One of 88 officially recognized areas in the sky. The term constellation also refers to the star pattern inside a given area. Constellation boundaries are analogous to state or provincial boundaries.

culmination: The instant when a celestial object reaches its highest point in the sky; i.e., when it is on the meridian.

declination (dec.): A parallel of latitude on the celestial sphere, equivalent to geographic latitude on the Earth. The declination of a sky object is expressed in degrees north or south of the celestial equator.

deep-sky object: Celestial objects located beyond the solar system. Deep-sky objects include double stars, star clusters, nebulas, and galaxies.

double star: An informal term for binary star. Most of the examples in this book are not true binary stars but wide "optical doubles" whose components happen to lie in the same line of sight as viewed from Earth.

eclipsing binary: Two stars of unequal brightness in a binary system that is oriented "edge-on" to our line-of-sight. The total light of the system decreases when the dim star passes in front of the bright star on each orbit.

ecliptic: The path the Sun follows annually in an eastward direction around the celestial sphere.

emission nebula: see nebula.

equinox: The Sun crosses the celestial equator, heading northward, at the *spring,* or *vernal, equinox* on or about March 21. The Sun crosses the equator, heading southward, at the *autumn equinox* around September 21. During an equinox, the Sun rises due east and sets due west; day and night are approximately equal in length.

galaxy: A large, rotating system of stars bound by gravity. Major spiral galaxies, such as the nearby Andromeda Galaxy and our own Milky Way Galaxy, feature a hub of mostly older suns surrounded by a flat, pinwheel-shaped disk of mostly younger stars. Galaxies also contain vast amounts of gas and dust.

giant: A large, highly luminous star. Giants represent a late phase in stellar evolution during which the star has greatly expanded. Supergiants are even larger and brighter than giants. Only the most massive stars can become supergiants. Betelgeuse in Orion is a red supergiant; Deneb in Cygnus is a blue supergiant.

globular cluster: A large, dense, gravitationally-bound ball of several hundred thousand extremely old stars. About 150 globular clusters are scattered in a spherical halo around the Milky Way Galaxy.

heliacal rising: The first appearance of a bright star low in the dawn sky after a period of invisibility due to its proximity to the Sun.

light-year: The distance that light travels in one year, which is nearly 6 trillion miles (9.5 trillion kilometers). The speed of light is approximately 186,000 miles per second (300,000 kilometers per second).

lucida: The brightest star of a constellation.

Magellanic Clouds: The two irregularly-shaped "dwarf" galaxies, which are satellites of our Milky Way Galaxy. The Large and Small Magellanic Clouds are located in the deep southern celestial hemisphere.

magnitude: A measurement of the brightness of a star or other celestial object. Magnitude 1 stars are among the brightest in the sky. Stars of magnitude 6 are at the limit of naked-eye visibility.

meridian: The line passing through the zenith that joins the north and south points on the horizon. The meridian divides the visible sky in half. Stars on the meridian have reached their greatest height (see **culmination**).

Messier object: Any of 110 bright clusters, nebulas, and galaxies in a catalog compiled by 18th-century French astronomer Charles Messier. The original list, published in 1774, contained 45 objects but has been supplemented with additional objects over the years.

Milky Way: The name given to the spiral galaxy in which our solar system is located. It is also the band of light, representing portions of our galaxy's spiral arms, that arches across the sky.

nebula: A cloud of interstellar gas and dust. An *emission nebula* (the type of nebula most often cited in this book) glows in the presence of high-energy radiation emanating from extremely hot stars embedded in the nebula.

open cluster: A loosely-bound collection of several hundred relatively young stars. Most open clusters reside in or near the plane of the Milky Way Galaxy.

Polaris: The 2nd-magnitude star Alpha (α) Ursae Minoris, located less than one degree from the north celestial pole. Also known as the North Star, Polaris is the brightest star in the Little Dipper and Ursa Minor.

precession: The gravity of the Sun and the Moon exerts a torque on the Earth, causing the planet to wobble slightly on its axis. The precessional wobble takes approximately 26,000 years to complete.

right ascension (RA): A great circle on the celestial sphere, equivalent of geographic longitude. The hours of right ascension progress in an eastward direction around the celestial sphere.

solstice: The Sun reaches its greatest northern declination of 23½° at the summer solstice on or about June 21. The summer solstice is the longest day of the year. The Sun reaches its greatest southern declination of –23½° at the winter solstice around December 21. For northerners, the winter solstice is the shortest day of the year.

star: A large, luminous sphere of gas generating energy in its hot core via the process of nuclear fusion.

supergiant: See **giant**.

variable star: Any star whose light output varies, whether irregularly or in a regular cycle. Some stars physically pulsate due to internal instabilities. See also **eclipsing binary**.

zenith: The point on the sky that is directly above the observer's head.

zodiac: The band of 12 constellations in which the Sun, Moon, and planets travel around the celestial sphere. (The Moon can sometimes appear slightly outside the zodiac.) The Sun's annual motion around the sky bisects the zodiac. See also **ecliptic**.

▶ Resources

There's no end to the helpful resources available to novice stargazers today. A huge variety of astronomy books cover every aspect of the science. Monthly and bi-monthly magazines can keep you up-to-date on celestial events, while a range of star atlases, from entry-level to advanced, exist to tempt those who want to explore the sky more deeply. There is even planetarium software for your computer. Speaking of computers, the number of astronomy websites online is simply mind-boggling.

Some helpful skywatching resources, organized by category, are listed below.

MAGAZINES

Night Sky (SkyTonight.com)
This relatively new bimonthly magazine has been designed especially for entry-level sky observers. It features easy-to-read charts, constellation descriptions, helpful observing tips, equipment evaluations, and more. The writing is always clear and nontechnical.

Sky and Telescope (SkyTonight.com)
This acclaimed astronomical monthly has a worldwide reputation for in-depth astronomical news coverage, thought-provoking science articles, attractive maps and star charts, and awe-inspiring astrophotography.

Skywatch (Sky Publishing, an annual publication)
An immensely useful compendium of sky information for the current year, this handy guide includes monthly charts, condensed, quick-find information on the visibility of the planets, and tips on buying and using a telescope. Available in August of the preceding year.

Sky News (www.skynewsmagazine.com)
Sky News is the Canadian magazine of astronomy and stargazing. In addition to charts tailored for observers north of latitude 45°, the magazine provides comprehensive descriptions of sky events, plus easy-to-read news updates.

ENTRY-LEVEL BOOKS ON STARGAZING & CONSTELLATION LORE

Night Watch:
A Practical Guide to Viewing the Universe, 4th Edition
Terence Dickinson (Firefly)
This is one of the finest introductions to the night sky ever written, expertly covering a range of observing activities with the naked eye, binoculars, and small telescopes.

The Backyard Stargazer:
An Absolute Beginner's Guide to Skywatching with and without a Telescope
Pat Price (Quarry Books)
An informal and highly informative guide to stargazing, this well-illustrated book touches on a variety of celestial topics. The book includes dozens of simple and informative observing projects to get you going in the hobby.

The Monthly Sky Guide
Ian Ridpath and Wil Tirion,
6th Edition (Cambridge)
An incisive month-by-month description of the night sky, this guide includes excellent charts, close-ups of interesting star regions, and notes on where to find the planets.

The New Patterns in the Sky:
Myths and Legends of the Stars
Julius D.W. Staal (McDonald & Woodward)
This in-depth treatment of constellation mythology features the starlore of many ancient cultures.

STAR ATLASES

Sky & Telescope's Star Wheel and Night Sky Star Wheel

(Sky Publishing)

This planisphere is a simplified star atlas that will show you which constellations are visible on any date and time. It's available for different latitudes. The smaller Night Sky Star Wheel is designed for use in light-polluted skies.

Bright Star Atlas

Wil Tirion and Brian Skiff

(Willmann-Bell)

One of the best naked-eye star guides ever created, these maps contain nearly 9,000 stars with magnitudes as dim as 6.5. More than 600 bright deep-sky objects are listed.

The Stars: A New Way to See Them

H.A. Rey

This informal guidebook to the constellations — a perennial favorite among novice stargazers — employs classy cartoon illustrations and detailed charts to "connect the dots" of the constellations in a novel way.

Starry Night (planetarium software)

www.starrynight.com

The Sky (planetarium software)

www.bisque.com

These exceptional software programs realistically depict the night sky, including thousands of stars, numerous deep-sky objects, and the positions of the Sun, Moon, and planets, for any time of night from any location on Earth. Planetarium software can demonstrate a year's worth of sky motion in a matter of seconds.

Pocket Sky Atlas

(Sky Publishing)

For those wanting to go beyond an entry-level star chart, consider this small 80-chart atlas containing more than 30,000 stars and 1,500 deep-sky objects.

HELPFUL WEBSITES

Cartes du Ciel (Sky Charts)

www.stargazing.net/astropc

A basic software star atlas, Cartes du Ciel can be downloaded to your computer for free.

Sky Publishing

SkyTonight.com

This site links to two magazines — the monthly *Sky & Telescope* and bimonthly *Night Sky* — and contains (among other things) observing articles, how-to tips, current sky events, an interactive sky chart, planetarium and astronomy club listings, and astronomy news. You'll also find predictions of Algol's "winks" (page 81) here: SkyTonight.com/observing/objects/javascript/3304096.html

Constellations

http://www.dibonsmith.com

This informative website covers all 88 constellations. It includes mythological summaries plus descriptions of key stars and interesting deep-sky objects.

ASTRONOMY CLUBS AND PLANETARIUMS

For worldwide listings of astronomy clubs, see:

SkyTonight.com/resources/organizations

www.lpl.arizona.edu/~rhill/alpo/clublinks/html

www.astronomyclubs.com

For listings of planetariums and science centers, see:

SkyTonight.com/resources/organizations

www.lochness.com/lpco/lpco.html

http://www.ips-planetarium.org/atw/ips-worldwide.html

▶ Index

Page numbers in **boldface** mean that the subject appears in a photograph, chart, or illustration.

▶ Image Credits

All star charts: Sky Publishing
All constellation icons: *The Star Atlas,* Johannes Hevelius
All illustrations: Sky Publishing
All photographs (except as noted below): Akira Fujii

page *vi* Alan Dyer
page 2 Todd Carlson (x2)
page 3 Canada-France-Hawaii Telescope / J.-C. Cuillandre / *Coelum*
page 4 [upper] Jeff Ball
page 5 NASA
page 8 [lower] Gemini Observatory / Peter Michaud
page 18 Ken Hewitt-White
page 25 Dennis di Cicco
page 33 [right] Martin C. Germano
page 53 P.K. Chen
page 68 [inset] Bob & Janice Fera
page 90 Dennis di Cicco

Summer Night Sky

When To Use This Star Map

Early June:	1 a.m.
Late June:	midnight
Early July:	11 p.m.
Late July:	10 p.m.
Early August:	9 p.m.
Late August:	Dusk

This star chart is most accurate if used within an hour or so of the times listed and is plotted for observers located between 30° and 50° north latitude. All times are standard time; if daylight-saving time is in effect, add one hour.

To use this chart, hold it in front of you and rotate it so that the yellow label corresponding to the direction you are facing is positioned at the bottom, right-side up. The stars in the sky should match those depicted on the chart. Ignore all the parts of the map above horizons you are not facing.

Autumn Night Sky

When To Use This Star Map

Early September:	midnight
Late September:	11 p.m.
Early October:	10 p.m.
Late October:	9 p.m.
Early November:	8 p.m.
Late November:	7 p.m.

This star chart is most accurate if used within an hour or so of the times listed and is plotted for observers located between 30° and 50° north latitude. All times are standard time; if daylight-saving time is in effect, add one hour.

To use this chart, hold it in front of you and rotate it so that the yellow label corresponding to the direction you are facing is positioned at the bottom, right-side up. The stars in the sky should match those depicted on the chart. The farther up from the map's edge they appear, the higher they'll be shining in your sky.

Star magnitudes
−1 0 1 2 3 4

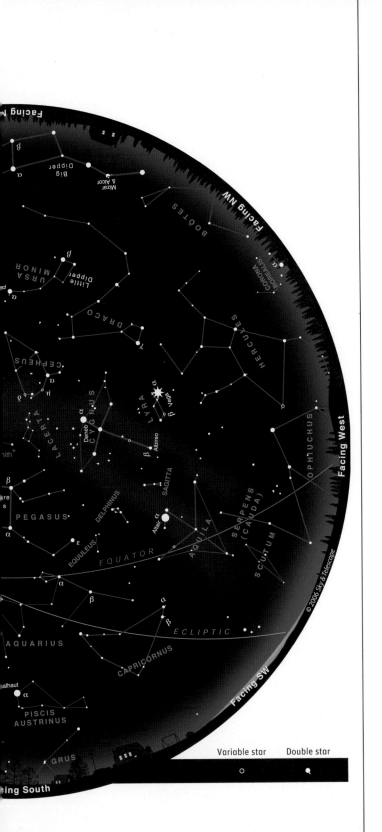

Variable star   Double star

○   ●

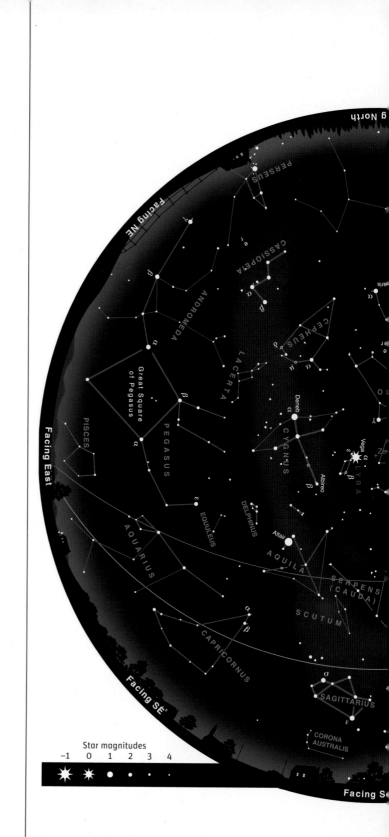

Facing North
Facing NE
Facing East
Facing SE'
Facing S

PERSEUS

CASSIOPEIA

ANDROMEDA

LACERTA

CEPHEUS

Great Square
of Pegasus

PEGASUS

PISCES

Deneb
α
CYGNUS

Vega
LYRA
β
Albireo

DELPHINUS

EQUULEUS

AQUARIUS

AQUILA
Altair

SERPENS
(CAUDA)

SCUTUM

CAPRICORNUS
α
β

SAGITTARIUS
σ

CORONA
AUSTRALIS

Star magnitudes

-1 0 1 2 3 4